INTERNATIONAL CENTRE FOR MECHANICAL SCIENCES

COURSES AND LECTURES - No. 60

GUIDO BELFORTE

TECHNICAL UNIVERSITY OF TURIN

FLUIDIC APPLICATIONS

COURSE HELD AT THE DEPARTMENT
OF HYDRO-AND GAS-DYNAMICS
OCTOBER 1970

UDINE 1973

SPRINGER-VERLAG WIEN GMBH

ISBN 978-3-211-81220-4   ISBN 978-3-7091-2852-7 (eBook)
DOI 10.1007/978-3-7091-2852-7

P R E F A C E

Fluidic elements and systems are studied
and applied since about ten years. Scientific and technical interest is proved by many international confer-
ences. In many University and Research Centres theorical aspects and complex fluid phenomena are studied.
Many firms, too, work in fluidic field, and applications increase. Some textbooks have been printed, in
particular, in last years.

But, "Fluidics", is a complex technology,
so some aspects have been examined only partially. In
my work, I wanted to describe and discuss some particular problems. In particular, coupling problems, medical and metrologic applications, power and interface
elements are considered.

I hope that this work may be useful to work-
ers in this field.

Guido Belforte

Udine, September I970

Chapter 1

FLUIDIC ELEMENT DESIGN

1. Introduction

The fluidic element design is the study of the single devices which compose a fluidic system.

This study aims at getting elements which display a given function in the desired way.

Therefore the fluidic element design includes the problem of sizing and optimisation of element geometry.

This aspect of the question requires therefore a good knowledge of the complex fluid dynamic phenomena which are the foundations of the element operation and of the parameters which these phenomena depend on [1].

Another fundamental aspect of design is however the following one : the fluidic element must not be considered as an insulated entity, but as a part of a more or less complicated circuit. One should therefore define its relations with other elements and determine its condition of operation when it is plugged in a circuit.

This is done depending on the operation characteristics [2] [3]. The latter are expressed with different curves depending on the kind of element (digital, proportional, wall effect, turbulence, ...).

The characteristics may be either of static or dynamic type, depending on the conditions of survey.

For most of applications it is enough to know the static characteristics (which will be considered later on).

Only in some particular cases (very high working frequencies, ...) it is necessary to use dynamic curves.

The most important operation characteristics are
those ones of wall effect digital elements (bistable and OR-NOR)
owing to the wide use of these elements.

2. Operation characteristics of wall effect digital elements

These curves are of three kinds :

- supply

- output

- control.

2.1. Supply characteristic

The supply characteristic is the curve which represents
the relationship between supply flow and supply pressure.

The typical trend of the curve is shown in Fig. 1,
where the abscissa is supply pressure and the ordinate is
supply flow. The curve gives the consumption when the operation pressure of the element is known. It should be noted that
the supply flow of air depends on supply pressure only and is
independent on output impedance.

Working pressures of commercial
wall effect elements vary from
0.05 to 1.5 $\dfrac{Kg.}{cm^2}$. The flows (which
depend on the cross section of ducts
and on supply pressures) vary bet-
ween 20 and 100 $\dfrac{cm^3}{sec}$. The power,
given by the product of pressure
times flow, is about $0.5 \div 1w$ for
wall effect elements and $0.1w$ for

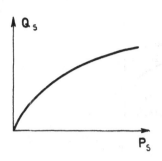

Fig. 1

turbulence elements whose supply pressure is much lower
$(0.015 \div 0.020 \dfrac{Kg}{cm^2})$.

However, it is possible to reach a power of 10mw with
turbulence amplifiers and a power of less than 100mw with wall
effect elements $[4]$.

2.2. Output characteristic

This curve gives the relationship between flow and
pressure in each output duct.

The curve is traced for a given value of supply pressure and is composed by two different branches (Fig. 2). The
upper branch describes the behaviour of the output duct when
it is active, that is when the power flow passes through it; the
lower branch describes the behaviour of the same output when
it is not active, that is when the power flow passes through the

other output. In these curves the air flow is considered positive when it leaves the element.

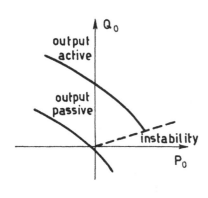

Fig. 2

When the output pressure increases the corresponding flow decreases. This matching of the element to output impedance is possible, even if supply flow is constant, if there are vents. An air flow, depending on output impedance, always passes through these vents to atmosphere.

If the element is sensitive to output impedance, the branch traced for the active output ends at the value of P_0 which makes the power jet instable on output duct. Otherwise, the curve goes on even below the P_0 axis.

When the supply pressure P_s changes, the curves also change showing a constant trend. When P_s increases, they move towards higher values of flow and pressure.

2.3. Control characteristic

This characteristic gives the relationship between control pressure and control flow and gives the values of both pressure and flow which are necessary for switching (Fig. 3). In these curves the flow is positive when enters into the element. Now let us switch the power jet by supplying air to the

corresponding control. A manometer
connected to the control duct measures the depression corresponding to
the mean value of the pressure in the
separation bubble (point A).

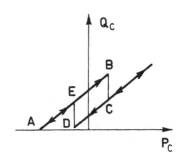

    By connecting the control to
a pressure regulator and increasing
the pressure, the control flow also

Fig. 3

increases (line AEB) up to point B
where the jet switches. As a consequence of switching C becomes the operation point.

    The entrainment due to the jet may now be different
as the control position does not correspond to the bubble position. If P_c still increases the flow increases again.

    If now the control pressure decreases, Q_c decreases
and point C is surpassed with no switching. Then, a new section of curve follows (CD) until the jet switches (point D). A
further decrease of control pressure gives again the curve EA.
The switching shows then a hysteresis : the control pressure
which allows the reattachment of the jet on its own side is lower than the pressure which causes the separation.

    For a given element and a given supply pressure the
control characteristic depends on the pressure value at the other control and on the output impedance. The effect of an increase of the pressure at the other control is shown in Fig. 4:

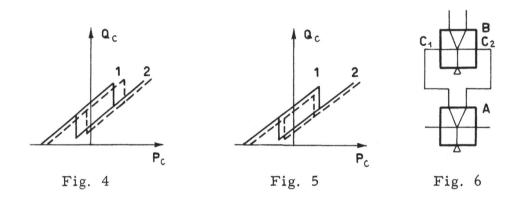

Fig. 4   Fig. 5   Fig. 6

the curve dips down and switching occurs at higher pressure and flow (curve 2). A similar phenomenon occurs in OR-NOR elements where pressure values P_B and P_D are both positive.

The increase of output impedance also causes a sinking of the curve, but switching occurs with flow and pressure lower than in the former case (Fig. 5).

An increase of supply pressure causes an increase of both switching flow and pressure.

The last parameter is a dynamic one, that is the rate of change of the control pressure.

3. Use of the characteristics in bistable coupling

By means of the former characteristic curves it is possible to forecast if bistable A of Fig. 6 can switch bistable B. One should superpose the output characteristic of the first bistable to the control characteristic of the second one and to inspect the intersection.

There are several cases :

1st case :

The trend of the curves is similar to that one shown in Fig. 7. The intersections are at least four. In this case the bistable A cannot control the bistable B.

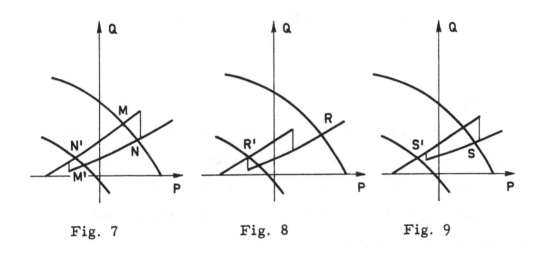

Fig. 7   Fig. 8   Fig. 9

As in any intersection point output pressure and flow of the controlling element (A) are equal to control pressure and flow of the controlled element (B), it follows that any intersection point can be a possible operation point. In the present case for each condition of bistable A (e. g. right output is active) two conditions are possible for the other element and these conditions correspond to the couples of points MM and NN'.

In the first case the right output of the second element
is active. The point M of control C_2 belongs in fact to the
branch of control characteristic which corresponds to the configuration with the jet attached to the same side of the examined control. The point M' gives the condition of operation on
control C_1 and belongs to the branch of the curve which represents the power jet attached to the wall opposite to the control.

In the second case (NN') the left output of bistable
is active.

2nd case

The upper branch of output characteristic has only one
point of intersection with control characteristic (Fig. 8). In
this case the element A controls B . Now, in fact, for each
condition of operation of A, B has only one possible condition of
operation, corresponding to points RR'.

If the active A output is the right one, the active B
output is the left one.

In fact the point R of control C_2 belongs to the lower branch of the control curve corresponding to the active left
output of B. In the same condition the operation point of the
left output is R'. In this case pressure and flow of the first
element are large enough to control the second one.

3rd case.

The lower branch of output characteristic has only one point of intersection with control characteristic (Fig. 9). This case is quite similar to the former one, as now there is only one couple of operation points SS' too.

Now, however, it is the passive branch which causes the switching of the controlled element, as it is the passive branch which has one point of intersection with the control curve.

In this case the control is a depression control while in the former case it was a pressure control.

4. Fan-out determination

We can now generalize what has been said to study the problem of controlling more than one element with one pilot element. In this case any controlled element needs its own control flow and pressure.

One should then add each other the control characteristics of the elements which have to be controlled and to repeat the former arguments on the resulting curve. This procedure is valid even when the controlled elements are different among themselves and therefore have different control curves. However, the most important case is when all the elements are equal to the piloting element. This is the case shown in Fig. 10 where the resulting control characteristics are determined with

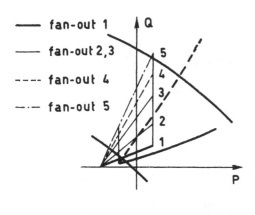

— fan-out 1

— fan-out 2,3

---- fan-out 4

—·— fan-out 5

Fig. 10

the number of elements connected in parallel. One can
see that the maximum number
of elements which are controlled is 4, as the curve 5 has
more than one point of intersection with the upper branch
of output characteristic, while
there are always two points of
intersection with the lower
branch.

This number of elements, called fan-out, is one of the
characteristic data of the element and gives its controlling abil-
ity when it is connected in a circuit.

REFERENCES

[1]   Romiti, A. : "Fluid Dynamics of Jet Amplifiers",
       CISM, Udine (1970);

[2]   Kirshner, J.M. : "Fluid Amplifiers", McGraw
       Hill, New York (1966), pag. 258 and fol.

[3]   Katz, S.; Dockery, R.J. : "Fluid Amplification",
       Number 11 : "Staging of Proportional
       and Bistable Fluid Amplifiers", Harry
       Diamond Laboratories, Washington D. C.,
       August 1963;

[4]   Chen, T.F. : "Low Power-Consumption Logic
       Devices" in "Fluidic Quarterly", vol. I,
       no. 2, Fluid Amplifier Associates, Ann
       Arbor, Michigan, January 1968.

LIST OF FIGURES

Chapter 2

MEDICAL APPLICATIONS

1. Introduction

Fluidic elements can find a useful employment in the field of medical equipment for many important causes :

1) the equipment must have a great reliability and fluidic elements can guarantee it;

2) the equipment must be fast and response time short. The required working frequencies and response times may be well obtained by fluidic equipment;

3) many applications are relative to fluid, both gases (air, oxygen, gas mixtures) and liquids (blood, water, urine). In some cases it is necessary to measure pressure or temperature, and to identify different gases. These operations may be performed by use of fluidic techniques.

Many applications have been made in fluid control : in the field of artificial respiration and of extracorporeal blood circulation.

Many kinds of artificial respirators have been built. Artificial respirators include either very simple respirators fit for emergency uses or more sophisticated ones which perform cycles, even of different shapes, which are useful in surgery and for clinical use.

It has been built also a model of membrane oxygenator
and several models of heart pump.

Many of these equipments have been already tested on
animals and men and not only on mechanical models.

Moreover, it has been built a sensor which can distinguish oxygen from nitrous oxide and avoid deadly accidents due
to an exchange of the two gases. Other applications are used in
artificial limb control .

Last but not least there are also external cardiac com-
pressors [1] and cardiac massage machines [2] .

2. Artificial respirators

2.1. Kinds of respirators

The purpose of artificial respirators is to supply a
proper ventilation to people whose natural respiration is no
longer sufficient. This is obtained by blowing air into the lung
pits where the air-blood contact occurs. During this operation
the blood is supplied with oxygen and the carbon dioxide is
carried away.

Artificial respiration, like natural one, can be divided
in two fundamental phases : inspiration, during which air enters
and is supplied to the lungs, and expiration, during which air,
now depleted of oxygen, is expelled from the lungs and new air
can enter again.

These two phases follow each other in a sequence of

cycles.

Fig. 1. shows the behaviour of the pressure during a cycle of period T.

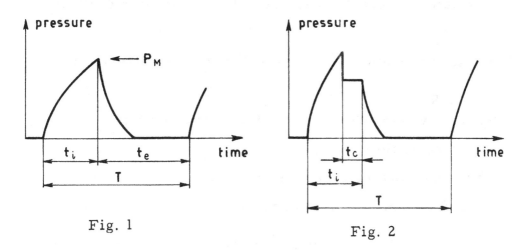

Fig. 1   Fig. 2

During the inspiration the pressure increases (time t_i), while during expiration it decreases down to zero (time t_e).

It is also possible to have cycles like the one shown in Fig. 2 where there is a constant pressure phase (t_c) during inspiration. This fact occurs when the supplying of air to the patient stops before the end of inspiration while the discharge is prevented. The air-oxygen exchange continues also during this constant pressure phase.

If one refers also to the adjustments which are possible in a respirator, the working paramenters during a cycle are :

1) cycle frequency f (or period $T = \frac{1}{f}$). It varies from

10-15 up to 50-60 cycle/min.

2) inspiration to expiration length ratio $R = \dfrac{t_i}{t_e}$. It varies from about $\dfrac{1}{3}$ to 1.

3) inspiration length t_i. By use of former parameters it is

$$t_i = \frac{R}{1+R} \cdot T.$$

4) expiration length t_e. By use of former parameters it is

$$t_e = \frac{T}{1+R} .$$

5) Supplied volume per cycle V. It can reach 1 litre.

6) Mean supply flow $V_m = V \cdot f$.

7) Maximum possible pressure P_M

8) Flux of supplied air Q. It depends on V and t_i.

$$V = \int_0^{t_i} Q \, dt.$$

Artificial respirators can be classified into three groups depending on which parameters have been chosen to set its operation :

1) Pressure cycled.

In respirators belonging to this group the inspiration stops when a given maximum pressure P_M is reached (Fig. 1).

The expiration stops when the pressure is decreased down to a minimum value. None of the other parameters (frequency, volume, etc.) can be fixed, and their value depends on the characteristics of lungs and aerial ducts of the patient.

2) Volume cycled.

The inspiration stops when a given volume of gas has
been supplied to the patient. The subsequent inspiration may
begin either when the pressure is decreased to a given value or
after a given time. Anyhow only the volume V is fixed. The
other parameters depend on the characteristics of the load applied to the respirator.

3) Time cycled.

This respirator offers the best possibilities of regulation. The operation is based on a timing element by means of
which it is possible to preset the cycle frequency. The respirators of this type are also volume cycled respirators. This can
be obtained in several ways.

The first one is to use an elastic capacitance (like bellows) to collect the air which must be supplied to the patient in
any cycle. The inspiration will then stop when the bellows is
empty or will continue for a time corresponding to a given value of R. In the latter case it is possible to have cycles like the
one shown in Fig. 2.

By using the elastic capacity it is possible to fix the
following parameters, also besides the frequency : inspiration-
expiration ratio   R (or t_i or t_e) and volume V (therefore,
the volume flow supplied).

In this case, of course, the supplied air flow must
satisfy to the condition that all the volume is supplied before

the end of respiration.

The second way to get a volume cycled operation is to have a constant air flow.

Then, if the inspiration length t_i is fixed, a given volume per cycle is obtained : $V = Q \cdot t_i$.

The mean flow which is supplied $V_m = \dfrac{Q\, t_i}{T}$ is proportional to the flow Q if R is constant.

Summarizing, with time-cycled respirators it is possible to fix the frequency f and the ratio R (or t_i and t_e) and another quantity, e. g. the volume V or the flow Q .

The other quantities depend on the former ones. One should also remeber that in no respirator the pressure must exceed a given maximum value to avoid dangerous damages to the patient lungs. Therefore any respirator has a safety valve set on this value.

Of course, when the valve is working, the operation is not volume cycled.

2.2. Pressure cycled respirators [1] [3] [4] [5] [6] [7] [8]

The operation of these respirators is controlled by the pressure level only.

From a fluidic point of view they can be built in a very simple way, by means of one element only. In any case the element is a bistable with one output connected to the patient while the other one discharges into atmosphere.

This respirator is quite fit for emergency use because it is very simple, has a small size, and its use is very easy.

Fig. 3. shows the scheme of such a respirator [1] [3].

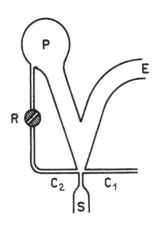

Output P is connected to the patient, while output E is the exhaust port. Control C_1 is connected to atmosphere, while a feedback duct and a resistance R connect C_2 to the patient duct. When the supply jet is switched to the left output, there is inspiration. Then the pressure in this branch increases until the pressure at the control C_2 is high enough to switch the jet.

Fig. 3

The expiration begins; the air from the lungs and supply air discharge through output E .

Therefore, the power jet entrains the vitiated air and this fact causes a depression in the duct P. When this depression has a sufficient value the power jet switches again to output P. A new cycle begins.

Curved wall elements [4] [5] [6] and elements without feedback channel [7] have been texted. In the latter case only the element geometry gives the needed sensivity to the load.

Fig. 4 shows the sketch of an emergency respirator

studied by Prof. A. Romiti and Ing. G. Belforte at the Poly-
technic of Turin.

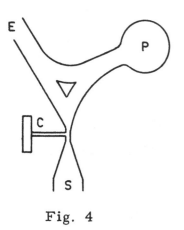

Fig. 4

The element has no feedback duct and has one control only on the side of output E. This control is connected to atmosphere by a variable port C. The element is provided with a duct which connects P to E; therefore, the air discharge from the patient is quite easier. Maximum (P_M) and minimum (P_m) pressure values, which correspond to jet switching, depend on geometry, feedback line resistance and possible polarizations of the controls.

For a given configuration they depend on supply pressure P_S (Fig. 5). Therefore, it is quite convenient to change P_S for changing P_M.

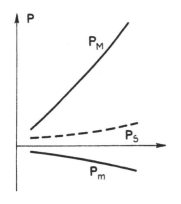

Fig. 5

Minimum pressure P_m are usually negative. It should be noticed that by closing more or less the valve of the element shown in Fig. 4 it is possible to get also positive minimum switching pressure (dashed line in Fig. 5).

2.3. Volume cycled respirators

Fig. 6 shows the sketch of a volume cycled respirator. In this respirator logic and power air, and the patient air have separated circuits, like the time cycles respirator shown in Fig. 8.

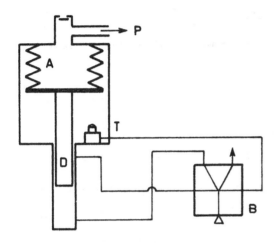

Fig. 6

The circuit separation can be useful when one must use anesthetics or other gas mixture in the patient circuit without possibility of gas diffusion in the external ambient. The respirator shown in Fig. 6 is formed by a bellows A which is compressed and expanded in an alternative way, thus supplying air to the patient P.

Some valves in the patient circuit arrange the operation in such a way that the inspiration occurs during the compression phase and expiration and filling of the bellows with new air occur during the expansion.

The bellows is operated by piston D controlled by bistable B.

When the left output of the bistable is active the phase in progress is inspiration and piston D moves upward. At the

end of the stroke a port which activates the feed-back line connected to the left control of bistable B is open. Then, the bistable   switches to the right output and the bellows moves
downward.

At the end of the descent the trigger T operates the
right control and a new cycle starts.

Another volume cycled respirator will be shown later
on (Figs. 8 and 9).

2.4. Time cycled respirators

In these respirators it is possible to control the frequency, the inspiration-expiration ratio and the patient air flow.
They can be constant flow devices or with elastic capacity. The
sketch of a constant flow respirator is shown in Fig. 7 $\begin{bmatrix} 8 \end{bmatrix}$.

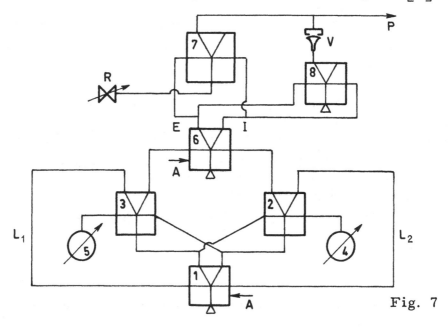

Fig. 7

The bistables 1, 2, 3 and the variable capacitances 4 and 5 form the timing system which is the base of the respirator operation.

The bistable 6, controlled by 2 and 3, has the right output I active during inspiration and the left one E active during expiration.

The operation of the timing system is as follows. Let us assume that the left output of the element 1 is active. The element 3 is then working and, as inside the capacitance 5 there is a certain pressure level (previous operation in the cycle), its active output is the right one (inspiration phase). At the same time the capacitance 4 is filled by the air which enters into the element 2 through the left control. Meanwhile the supply jet of the element 3 entrains the air contained in 5. As the capacitance 5 is closed its pressure decreases and will become a depression, then switching the power jet to the left output. The feed-back line L_1 is then active and switches the bistable 1 to the right output.

At this moment the expiration begins and during this phase the bistables 2 and 3 repeat the former cycle after exchanging their functions.

The presence of signal I at the right output of 6 causes the inspiration because the element 7 in this case supplies air to the patient P.

The supplied air flow is adjusted with the variable

resistance R.

The valve V, which is closed during the inspiration, allows a quick discharge of the vitiated air and reduces to a minimum its mixing with the fresh air which will be supplied during the next cycle.

If during the operation the maximum possible pressure is exceeded, a signal A, coming from a proper safety valve, switches the bistables 1 and 6.

Constant flow respirators, which control the flow by means of a system using a vortex valve and an impact modulator, have been studied too [10].

Fig. 8 (see next page) shows the sketch of a time cycled respirator with elastic capacity [11].

It should be noticed that in this respirator the value of frequency f and of ratio R are fixed, and not inspiration (t_i) and expiration (t_e) times like in the former ones.

The respirator has the following characteristics :

1) it is able to perform both automatic and assisted respiration with start cycle signal given by the patient itself;

2) cycle frequency is adjustable between 10 and 50 c.p.m. If a given frequency value is fixed, it remains constant and absolutely independent of the output impedance as it is given by an oscillator of the logic circuit;

3) the inspiration-period ratio is easily and safely variable between 0.25 and 0.50 with five different values.

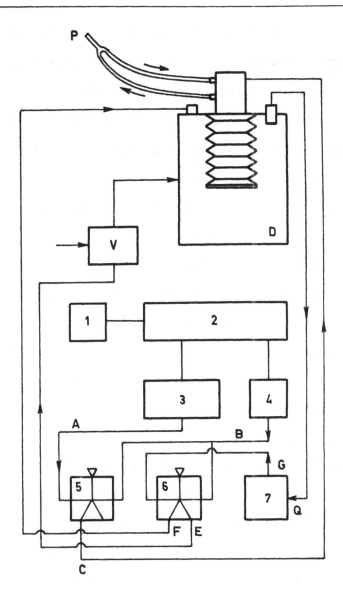

Fig. 8

Fig. 8 shows the block diagram of the logic circuit with automatic operation. An oscillator (1) with adjustable frequency transmits its pulses to a chain of binary counters (2). A series of elements (3) decodes the chain pulses and gives output pulses at different points of the cycle. Element 4 gives a B pulse at the beginning of each cycle.

At that moment, B pulse switches the flow in bistables 5 and 6 to C and F outputs.

C signal operates a valve placed on the patient circuit, blocking air discharge. F signal closes a dump valve placed on a bell D inside which there is a bellows containing the air for the patient. At the same time the absence of E signal acts a normally open valve V which blows air into the bell. The elastic capacity is then compressed and its content is given to the patient.

When the capacity is empty, a signal Q coming from an end of stroke sensor acts trigger 7 which switches bistable 6, allowing bell emptying and elastic capacity refilling.

This end of stroke signal was found necessary because of the capacity inertia in order to have as much time as possible to fill the capacity itself.

Opening of the valve placed on the patient circuit will take place when an A signal reaches bistable 5. At that moment expiration begins. Phase shifting between the end of elastic capacity emptying and valve opening assures the presence of a

Fig. 8a

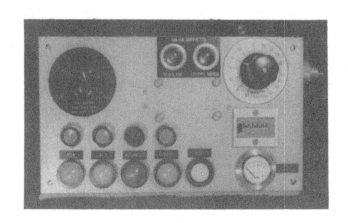

Fig. 10a

Fig. 10b

constant pressure phase inside the patient lungs.

Expiration beginning adjustment is made by alternative supply for some decoding elements by means of a switch.

For assisted cycle operation the diagram is very simplified (Fig. 9).

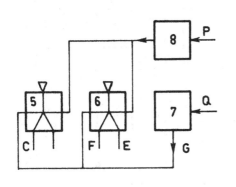

Fig. 9

In this case the beginning signal of each cycle is given by the patient itself by means of depression sensor 8. The depression generated by the patient is in fact transformed into a signal which switches bistables 5 and 6 to C and F outputs, beginning to blow in air. By means of trigger 7 the end of stroke signal (Q) resets 5 and 6 that are therefore ready for the next cycle.

The detailed diagram of the fluidic circuit used for automatic cycle operation is shown in Fig. 10 (see next page).

The oscillator 9 controls the binary counter cascade (12, 13, 14, 15). An OR-NOR (11) is necessary in order to regenerate the oscillator signals before sending them to the binary counters. In this way a safe operation is ensured.

An induction oscillator is used. The oscillator is a bistable without vents, whose outputs are connected to OR-NOR

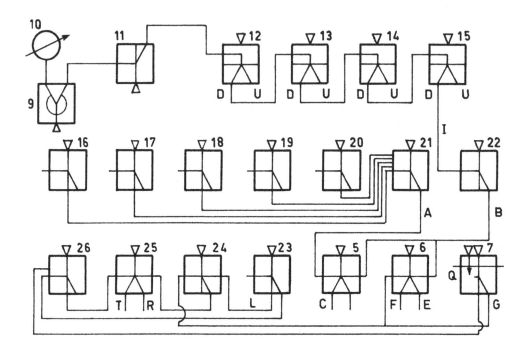

Fig. 10

11 and to a variable capacity 10 respectively. With this oscillator it is possible to get low frequencies with a good regularity.

The two different counter states have been labelled
with U (UP) and D (DOWN). As the counters are four, 16
oscillator pulses are needed to make a full cycle.

OR-NOR 16, 17, 18, 19, 20 are the decoding elements
to obtain the signal A of the expiration start. Cycle fractions
that have been carried out and corresponding input signals are:
4/16, 12U, 14D; 5/16, 12D, 14D; 6/16, 12U, 13D, 14D; 7/16,
12D, 13D, 14D; 8/16, 13U, 15D.

The pulse limiter 22 gives the start cycle signal. It is controlled by the signal 15D.

The circuit formed by elements 23, 24, 25, 26 signals if the elastic capacity emptying is ended before or after 2/16 of the cycle. In other words, it says if the phase during which there is air flow towards the patient is ended before or after 2/16 of the cycle.

OR-NOR 23, controlled by signals 13U, 14U, 15U, gives a signal L which lasts for the first 2/16 of the cycle. If the end of delivery signal G arrives after this period, the inhibited OR-NOR 24 switches bistable 25 to the position T. If signal G arrives while L is still present, OR-NOR 26 is switched, therefore switching the bistable 25 to the position R.

Two pilot lamps, connected to T and R, allow observers outside the control box to know which one of the two alternatives has taken place.

Two sensors have been used to complete the control system : the first one signals the end of elastic capacity emptying; the second one gives the start assisted operation cycle signal.

This is used to get, from a depression signal, a positive pressure able to control fluidic elements. It has been made (Fig. 11) with a bistable controlled by a mushroom elastic valve A connected to the patient P.

This valve closes the right control port when internal

pressure is atmospheric. In such conditions the active output
is **Y**. If inside the valve there is a depression, the valve con-
tracts connecting the control to atmosphere. In such conditions
the active output is **W**.

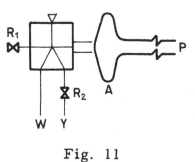

Fig. 11

The element is now monostable owing
to resistance R_1 and R_2, to output
W impedance and the control resistance. Two adjustment screws allow a
variation of the distance between mushroom valve and control port, thus
changing the sensitivity. With this
sensor a switching pressure of -5 mm $_{H_2O}$ was reached. This
is a more than enough value in the case of start cycle signal
given by the patient, and not the apparatus limit.

In fact the switching pressure may be positive too.

3. Artificial heart pump [1] [12] [13]

The artificial heart is used to ensure an auxiliary
blood circulation. It can be used during heart surgery or to
assist the natural circulation during crises or diseases.

The total and permanent replacement of the natural
heart by an artificial one seems to be still far away in the fu-
ture.

During the assistance to natural heart, the artificial
heart pump can be connected in series or in parallel. The prob-

lems joined to the artificial heart use are mainly due to the difficulty of handling the blood without damaging it. These difficulties increase and multiply with time for a given equipment in such a way that it is not possible to use an equipment of this kind beyond a given time limit.

The following points must be respected :

1) thrombus formation, i.e. blood coagulation, must be avoided.

2) the blood components must be damaged as little as possible.

To get this shocks and too high pressures must be avoided.

3) artificial and natural heart must have similar wave shapes.

Besides this the artificial heart pump must be a quite reliable equipment and it must give the possibility of setting frequency, pressure and volume of pumped blood per cycle.

Fluidics can fulfil very well some of the former conditions.

The use of static elements leads to build equipments with a minimum quantity of moving members and therefore more reliable than other ones.

One should also bear in mind that owing to the sensitivity of fluidic elements to the load the blood flow rate changes in an automatic way, avoiding blood damages.

Some tests have shown that fluidic artificial heart pumps can be used for a longer time than conventional roller pumps [12].

Fig. 12 shows the sketch of an artificial heart pump controlled by a bistable [1].

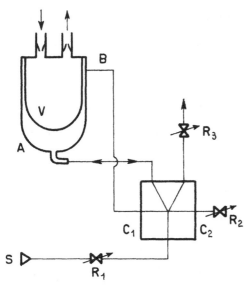

The blood is collected in an artificial ventricle V formed by an elastic bag, which is contained in a rigid chamber A. The ventricle is connected to the blood circuit by means of an input and an output duct. In these two ducts there are some nonreturn valves which direct the blood in the correct way. A bistable provides the ventricle control.

Fig. 12

When the left output of the bistable is active the air enters into the chamber and compresses the ventricle pushing the blood towards the output. When the contraction reaches the desired value the bag opens a port B which connects the active output to the control C_1. Then, the bistable switches. The right output is now active and discharges into the atmosphere both the supply air and the air contained in A through the resistance R_3.

In such a way the chamber is emptied and the ventricle is filled again.

When the ventricle is filled the port B is closed and

the bistable switches again to the left output therefore beginning a new cycle.

The adjustment is made by means of the variable resistances R_1; R_2; R_3.

The resistance R_1 which controls the supply pressure of the bistable, and therefore the flow which enters into the chamber A , works like a pulse amplitude control valve. Therefore R_1 controls also the pressure obtained in the chamber and the thrust on the ventricle.

The resistance R_2 controls the air entrained through control C_2. This resistance is also called pulse duration control valve and sets the filling time.

The resistance R_3 (pulse rate control valve) controls the flow which discharges into atmosphere from the right output and therefore the pressure decrease rate in chamber A .

This pump has reached a flow of 10 1/min. , maximum pressure of 500 mm$_{Hg}$ and frequencies of 30-130 cycle/min.

It should be noticed also that the blood flow increases when the blood supply pressure increases and decreases when the load impedances increase.

Fig. 13 shows the sketch of another artificial heart pump [12] . The pump is now formed by a piston with two rolling membranes M made with silicon rubber which separate two chambers. The upper chamber A is filled and emptied with blood.

Like in the former case there are two ducts with non-

return valves. These

valves are of the leaflet

type. The lower chamber

B is filled and emptied

with air in an alternative

way to pump the blood.

The control is obtained

by using a bistable, whose

left output forms a feed-

back loop, through the

chamber B , with the

left control duct.

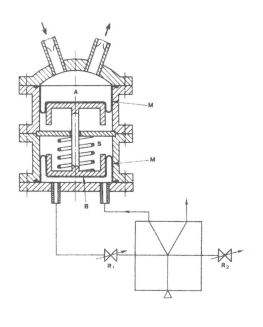

Fig. 13

When this output is ac-
tive, the pressure in the chamber B increases until the jet
switches. Then, the chamber is emptied, and the emptying
ends, when the pressure is low enough the jet becomes unstable
and switches to the left output.

The adjustment is made by setting the values of resis
tances R_1 and R_2 which allow to change the values of pressure
and flow of the switching points of the bistable. In such a way
it is possible to change the frequency and the volume supplied
during one cycle by changing the piston stroke.

In order to avoid shocks due to a sudden stop of the
piston a spring S and elastic stops are used.

This kind of pump has been used for tests "in vivo" on dogs.

Last but not least, artificial heart pumps controlled by a system formed with two elements, a bistable and a monostable connected each other with delay lines, have been built [13]

4. Membrane oxygenator [14]

The purpose of this equipment, used in surgery, is to give oxygen to the blood and to remove the carbon dioxide. This exchange occurs through a semipermeable membrane.

The sketch of the con trol part of the equipment is shown in Fig. 14. The oxygen supplied to the membrane M is controlled by a spool valve. The alternative movement of the spool valve is obtained with a bistable. The connection of the

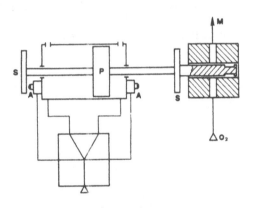

Fig. 14

spool valve to the piston P is rigid. The air which flows through one of the output ducts of the bistable causes a displacement of the piston. When the piston is at the end of its stroke, the striker S operates the trigger A and lets the air flow towards the control, therefore switching the element.
After the switching, the piston begins the reverse motion.

5. Gas type detector [15]

The purpose of the equipment is to check the kind of gas which flows in a pipe or is contained in a tank. A particular and interesting case is to distinguish the oxygen O_2 from the nitrous oxide N_2O.

Both gases are used in anesthesia and they must not be exchanged. The two gases are distinguished in a very simple way by using a fluidic element whose operation is based on laminar-turbulent transition.

The sketch of this element, which is also used as fluidic amplifier (step-pipe amplifier) is shown in Fig. 15.

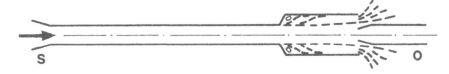

Fig. 15

The supply flow issues from the end of a long tube into a chamber whose cross-section is quite larger than the one of the tube and open to atmosphere.

A collector 0 collects the output flow.

If the jet is laminar the recovered pressure is high, while when the jet is turbulent it attaches to the chamber wall (Coanda effect) and discharges into atmosphere. The output pressure then decreases suddenly.

In Fig. 16 the pressure P_0 is plotted versus Reynolds

number R_e . The effect of transition is clearly shown. It is
shown also the transition hysteresis.

The critical Reynolds number is about 3000.

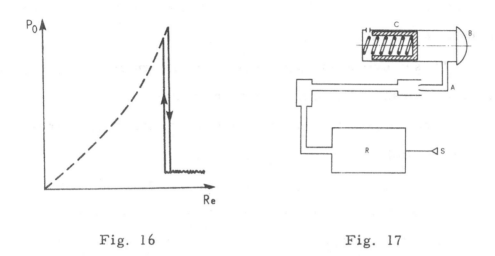

Fig. 16   Fig. 17

A step-pipe amplifier of this kind (**A**) is used in
the gas type detector (Fig. 17) and is supplied with a constant
pressure by means of a pressure regulator **R** .

The equipment is set in such a way that Reynolds num-
bers correspondent to oxygen and to nitrous oxide are respectively lower and higher than the critical value.

This is possible because the kinematic viscosities of
the two gases are quite different (O_2 viscosity is about twice
the NO_2 viscosity).

The element output is connected to the fluidic lamp **B**.

The oxygen jet is laminar and the high recovered pres-

sure pushes the small piston C to the left side. The NO_2 jet
is turbulent and the spring pushes the piston C to the right
side. Therefore, the piston position reveals the kind of gas
flowing in the pipe.

6. Prosthetic devices [16] [17]

These devices are used when a limb is missing or
when muscles cannot be stimulated with electricity.

In such cases it is necessary to have artificial extern-
ally powered limbs.

A pneumatic technique using small actuator pistons and
carbon dioxide as power fluid seems quite fit for the case.

The pneumatic pistons can be controlled by means of
0-Ring valves and the input signal is given by means of a small
lever.

A valve of this kind is shown in Fig. 18.

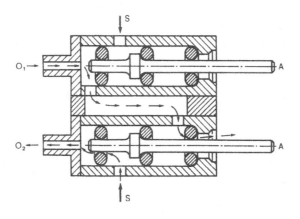

Fig. 18

Its operation is based on the following facts.

Let us introduce a spindle in an 0-Ring placed in its housing. The 0-Ring separates two ambients with different pressures.

If the spindle centered with the 0-Ring there are no leaks.

If one gives to the spindle a radial displacement there is a flow of fluid. The larger the displacement, the larger the flow.

The valve shown in Fig. 18 has two spindles A which can be activated in an alternative way.

The valve is connected to the two chambers of the cylinder (0_1 ; 0_2), to the supply S and its own chambers can be connected to the external ambient to exhaust the gas issuing from the cylinder.

If no spindle is activated the valve is in a neutral position and all the 0-Ring make leakproof contacts.

Activating a spindle causes a rotation around the central 0-Ring, which still is leakproof.

The two 0-Ring at the sides are no longer leakproof because the radial displacement is larger.

A chamber of the cylinder is then connected to supply S and the other one to exhaust. Activating the other spindle give reverse motion.

The three-way 0-Ring valve used as low pressure-to-high pressure interface (Fig. 20 ; Interface elements), also

can be used to control cylinders in artificial limbs.

REFERENCES

[1] Woodward, K.; Mon, G.; Joyce, J.; Straub; H.; Barila, T. : "Four Fluid Amplifier Controlled Medical Devices", Fluid Amplification Symposium, 1964, Harry Diamond Laboratories, Washington D. C.;

[2] Tsuchiya, K.; Kabei, N. : "Cardiac massage machine using fluidics", 4th Japanese Fluidics Symposium, 1969. Society of Instrument and Control Engineers, paper 37 (in japanese);

[3] Straub, H. H.; Meier, J. : "An evaluation of a fluid amplifier, face mask respirator". Fluid Amplification Symposium. October 1965. Harry Diamond Laboratories. Washington D. C., vol. III;

[4] Pavlin, C.; Kadosh, M. : "Mechanical characteristics of a pure fluid respirator with curved walls", 1st Cranfield Fluidics Conference. September 1965. British Hydromechanics Research Association. Cranfield, England, paper F3;

[5] Société Berlin : "Device for alternately filling and emptying an enclosure". British Patent. n. 1, 112, 365, May 1968;

[6] Chirana, Zavody Zdravotnickej Techniky, Odborovy Podnik (Czechoslovakia) : "Improvements in or relating to respirators". British Patent, 1,132,507. November 1968;

[7] Retec. Inc. (U.S.A.) : "Fluid flow switching device". British Patent 1,107,268, March 1968;

[8]   Ball, G. J. : "Application of Fluidics to automatic
        lung ventilation". 4th Cranfield Fluidics Confer-
        ence. March 1970. British Hydromechanics
        Research Association. Cranfield, England,
        paper M2;

[9]   Straub, H. H. : "Design requirements and proposal
        for Army Respirator", Report TR 1249, 1964.
        Harry Diamond Laboratories. Washington D. C. ;

[10]   Vince, J. R. ; Brown, C. C. : "The application of
        fluid jet devices to a medical respirator", 1st
        Cranfield Fluidics Conference. September
        1965. British Hydromechanics Research Asso-
        ciation. Cranfield, England, paper F4;

[11]   Belforte, G. : "Logic circuit of Artificial Respira-
        tor", 4th Cranfield Fluidics Conference. March
        1970. British Hydromechanics Research Asso-
        ciation. Cranfield, England, paper M1;

[12]   Tsuchiya, T. ; Nagakawa, K. ; Matsuda, Y. ; Miura,
        I. ; Niibori S. ; Yokosuka, T. : "Artificial Heart
        controlled by Fluid Amplifier", I. F. A. C. Sym-
        posium on Fluidics. November 1968, C. Baldwin
        Ltd. Tunbridge Wells, Kent, England, paper
        C7;

[13]   Harada, M. ; Ozaki, S. ; Hara, Y. : "Output charac-
        teristics of wall attachment elements", 3rd
        Cranfield Fluidics Conference, May 1968,
        British Hydromechanics Research Association,
        Cranfield, England, paper F2;

[14]   Mon, G. : "Design and Development of a Membrane
        Oxygenator", TR 1251, Harry Diamond Lab. ,
        Washington D. C. , 1964;

[15]   Rolland, J. N. : "L'amplificateur fluidique à évase-
        ment brusque", Automatisme, no. 5, mai 1969;

[16]   Kiessling, E.A. : "Pneumatic posthetic components
         rigid servo-mechanisms and their functional
         control". Application of External Power in
         Prosthetics and Orthotics, 1960, Publication
         874. National Academy of Sciences, National
         Research Council, Washington D.C.;

[17]   Lichtarowicz, A. : "0-Ring seal as a valve element".
         3rd Cranfield Fluidics Conference, May 1968,
         British Hydromechanics Research Association,
         Cranfield, England, paper E1.

LIST OF FIGURES

Chapter 3

METROLOGIC APPLICATIONS

1. Introduction

It is possible to utilize fluidic elements to measure some quantities like velocity, acceleration, temperature.

In such a way one gets instruments whose outputs are pneumatic signals.

These elements can be considered as interface elements between the quantities to be measured and the subsequent fluidic elements.

Their use is justified by several reasons, like simplicity and stoutness (fluidic anemometer), the presence of high temperatures or strong vibrations, the possibility of eliminating interface elements as these fluidic sensors can directly control pneumatic systems.

They have been built fluidic systems to measure the velocity of a fluid current, the angular velocity and the acceleration of body (even with inertial instruments), the temperature of a fluid and the density of a fluid.

2. Fluidic anemometer [1] [2] [3] [4]

It is used to measure the velocity of a fluid. The anemometer (see Fig. 1) is formed by a jet of fluid issuing from a supply duct S and by two collectors.

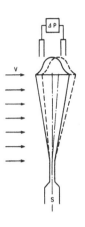

Fig. 1

Before reaching the collectors the jet must cross a rather long free space. If the surrounding fluid has a velocity V , like showed in Fig. 1, the jet will be deviated. The figure shows the velocity profiles of the jet at collector inputs. If there is no deviation (solid line profile), the same pressure is measured in the two outputs. If there is a deviation, as an effect of velocity V the collected pressures will be different.

The differential pressure Δ P is therefore depending on the velocity V .

These instruments are used in meteorology to measure wind velocity and in oceanography to measure sea currents [3]. The fluidic anemometer can measure very low velocities.

Such low velocities can be measured also with hot wires, which are too delicate and brittle for use in a rough environment while the fluidic anemometer can be a rugged instrument. Moreover, the fluidic anemometer should give a good response even if velocity varies with rather high frequency. It should be used coupled with pressure transducers and recorders.

It can be built with more than two collectors to measure also other velocity components.

The sensivity of commercial anemometers is about
$50 \, \dfrac{mm \, H_2O}{m/sec}$ in air and they work till 3 m/sec about. In water
the sensitivity is about $4100 \, \dfrac{mm \, H_2O}{m/sec}$ and working field till
0. 3 m/sec.

Fig. 2 shows another velocity sensor : the coflowing
sensor [1]. In this case the jet is parallel to the fluid current
that has to be investigated and has the same direction (upper
jet).

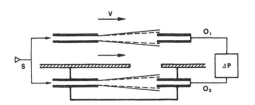

Fig. 2

Now a jet which crosses an
ambient which contains a
fluid entrains a certain a-
mount of the ambient fluid.
This amount depends on
many parameters (velocity,
kind of fluids, pressure,
and so forth).

Using two jets, one of them is placed in the ambient
with moving fluid (upper jet) and the other one is placed in a
still ambient with the same static pressure (lower jet), the en-
trainment effect is then different for the two jets only for the
part due to the velocity V. Then, the differential pressure
Δ P measured at the two outputs is proportional to velocity V.

The two jets are supplied with the same pressure to
have a sensor independent of supply pressure.

3. Measurement of angular velocity

Such measurements may be made to know the angular
velocity of a shaft, or of the body which the sensor is fixed to
(inertial instruments).

The first kind of measuring equipment may be applied
to turbine control, while the second kind equipments are useful
in missile stabilizing systems.

Fig. 3 shows the scheme of a digital method to measure the velocity [5]

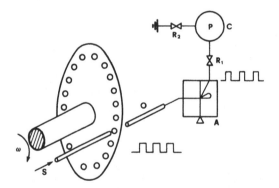

Fig. 3

The jet issuing from supply S is intercepted by a rotating drilled disc, in such a way that the collector tube O collects a series of pulses with frequency proportional to the angular velocity. These signals control a "one shot" A which

gives output pulses having constant amplitude.

The pulses are then integrated by a filter formed with resistances R_1 and R_2 , and capacitor C. The pressure P inside the filter is then proportional to the angular velocity. The frequency can be utilized in other ways to read the velocity value (e. g. : see later on the temperature sensitive oscillators).

Fig. 4 shows a boundary layer sensor [6]. It is a proportional device and gives an ouput signal which is proportional to the velocity.

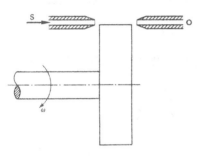

Fig. 4

The jet issuing from supply S is collected by output duct O. The jet itself passes through the boundary layer of the lateral surface of a rotating cylinder. The rotation of the disc generates a deviation of the jet which is proportional to the velocity. Of course, the larger the deviation, the lower the pressure recovered by O.

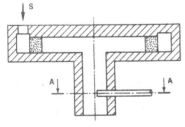

Fig. 5

Sec. A-A

Fig. 5 shows the sketch of a vortex rate sensor. This element belongs to the second kind equipments [4] [5] [7] [8] [9] [10] [11] [12] [13]. Its operation principle is based on the vortex effect. The air issuing from supply S enters into a vortex cham-

ber after passing through a porous ring R exhausting to atmosphere at the end of a cylindrical duct.

In the cylindrical duct are placed two aerodynamic
pickoffs whose position is shown in section AA (Fig. 5).

If the system does not rotate the flow in the vortex
chamber is radial, then it is axial in the cylindrical duct. The
two probes measure the same pressure.

If the system, and therefore the vortex element, rotates the flow receives the inertial rotation of the sensor while
it crosses the porous ring. The air gets therefore a tangential
velocity component and a vortex is generated.

This tangential component received by the inertial coupling element is amplified while it goes towards the center of the
chamber and is present in the output duct too. The two probes
(Fig. 5; sect. AA) measure then a differential pressure propor-
tional to the velocity. The sensivity is about $7 \frac{mm \, H_2O}{deg/sec}$ [12]. Of
course the sensivity depends on probe position and shape [7].

It is also possible to build vortex rate sensors where
a tube-type pickoff measures the vortex intensity.

The output duct looks like the one of a vented vortex
valve. The operation is based on the coning effect of the vortex
at the output : the probe collects a flow, and therefore a pressure, whose value depends on the angle of the cone formed by
the flow at the output of the element. Vortex rate sensors have

worked with temperatures up to 700° C $\begin{bmatrix} 11 \end{bmatrix}$.

4. Acceleration measurement

The fluidic systems can be utilized for acceleration measurements too.

Fig. 6 shows the scheme of a system fit for measuring linear accelerations along one direction $\begin{bmatrix} 14 \end{bmatrix}$.

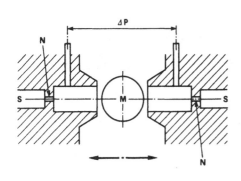

Fig. 6

The accelerometer is formed by a spherical seismic mass M balanced by two jets of air. The two jets are issuing from two opposed ducts which are supplied by supply S through narrow nozzles N. The sensitivity axis of the instrument is the duct and nozzle axis. When an acceleration parallel to this axis is present, the seismic mass is pushed against one of the ducts.

Then, the exhaust resistance of the corresponding jet increases, while that one of the other jet decreases.

A pressure difference ΔP rises between the two ducts and it can be detected by means of two small output ducts.

5. Temperature measurements

5.1. Introduction

Temperature can be measured in several different manners by using fluidic elements.

In fact, it should be remembered that the values of signal propagation velocity and of some fundamental parameters of fluidic circuits (resistances and capacitances), vary with temperature.

Moreover, other parameters affect the behaviour of elements and loops, say for instance supply pressure, kind of fluid, change of component size.

One must then build systems sensitive in some way to temperature changes and as insensitive as possible to the variations of other parameters and to external disturbances.

Several types of temperature sensors have been experimented. There exist several oscillators sensitive to temperature and capillary tube sensors.

There are then elements with elastic moving parts in which a change of temperature generates the deformation of a mechanical member. This deformation generates the output pressure variation.

5.2. Temperature sensitive oscillator [15] [16] [17] [18] [19]   )].

Each fluidic oscillator can be a temperature sensor. In fact it has been already said that the values of resistances,

capacities, and signal propagation velocity vary with temperature.

      Various kinds of oscillators are fit for this purpose :
feedback oscillator $\left[17\right]$, sonic oscillator $\left[15\right]$, edgetone.

      A sketch of a feedback oscillator is shown in Fig. 7.
It is essentially a bistable : feedback is obtained by connecting
outputs to controls with ducts F_1 and F_2 .

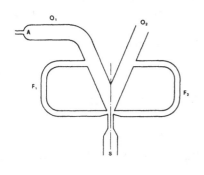

      In these elements the signal propa-
gates with a speed which is very
near the speed of sound. If one neg-
lects the switching time of the element with respect to the signal propagation time the frequency is given
by the relationship :

$$f = \frac{K \sqrt{T}}{\ell}$$

Fig. 7

      where K is a constant, T the ab-
solute temperature and ℓ is the feedback line length.

      The frequency is then proportional to the square root
of absolute temperature.

      Of course, these oscillators are also sensitive to oth-
er parameters and the most important one is the supply pressure. To make oscillators insensitive to supply pressure var-
iation one could use resonating ducts $\left[16\right]$ $\left[17\right]$. This is done
on the left output of the oscillator shown in Fig. 7. The output

duct O_1 ends with a very narrow port **A**, such that output duct
behaves like a closed resonator path.

Feedback oscillators can be made with proportional
elements too.

The sonic oscillator is formed by a bistable with control ducts connected each other by a single line.

When the jet switches to an output, from the two ends
of the line (i.e. from each control), a pressure wave and a
rarefaction wave start propagating in opposite directions in the
line. When each wave reaches the other control there is an un-
balance of the jet which switches.

Then, the phenomenon is repeated and one gets oscil-
lations.

The edgetone oscillator (see Fig. 8) is mainly formed
by a chamber partially divided in two equal parts by splitter
and having two outputs and one input (or supply).

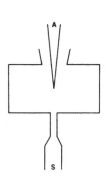

The jet issuing from the supply nozzle
strikes the splitter. Then, there is an alternate separation of vortexes from the
jet; these vortexes cause an alternate displacement of the jet therefore causing alternate filling and emptying of the two half
chambers. Frequencies are rather high
and depend on temperature and on the
chamber volume.

Fig. 8

All these elements can be useful to measure the temperature of a fluid One should supply the oscillator with the
fluid at unknown temperature.

In such a case, the response to a change of temperature is given when the new fluid has replaced the old one.

Moreover, if one wants to measure the temperature
in an ambient, he must place the oscillator or a part of its circuit in this ambient.

The operation is then more complicated because the
heat is transmitted from ambient to oscillator tubing and from
tubing to inner fluid.

With feedback oscillators it is possible to reach a sensitivity higher than 1 $\frac{Hz}{°C}$. Working temperature up to 1100° C
[17] [20] . The output signal of the sensor is therefore characterized by a variable frequency. There are many possibilities
for its utilization.

The first method is to use pressure transducers. Then
the electric signal is used.

Another method is to transform the frequency into an
analogue pressure signal. This is very useful if the remaining
part of the system is pneumatic.

In this case one may use a pulse limiter system with
integrating filter, as seen for the digital velocity sensor.

Anyway, the method which gives the best results is a

decoding technique with beats.

The scheme of such a circuit is shown in Fig. 9.

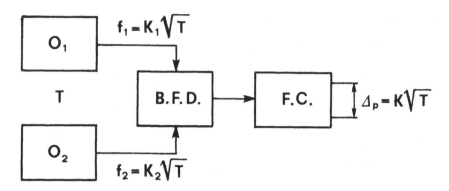

Fig. 9

Two oscillators O_1 and O_2 having different sensitivities are at the temperature T. Output signals which have respectively the frequencies $f_1 = K_1 \sqrt{T}$ and $f_2 = K_2 \sqrt{T}$ are in-
troduced into a beat frequency detector (B. F. D.) which generates the beats and, as final output, a signal whose frequency is
$\Delta f = f_1 - f_2 = (K_1 - K_2) \sqrt{T}$. This signal feeds the frequency
converter (F. C.) which transforms it into a pressure proportional to the frequency $f_1 - f_2$, then the temperature T [20].

The use of beats allows the system to increase its
sensitivity. In fact, working frequencies are high (hundreds or
thousands of Hz) and their variations due to the temperature
are small. By subtracting two frequencies of the same order of
magnitude one gets the low variation frequency of the temperature. Instead of using two oscillators with different sensitivi-

ties, it is also possible to use a temperature sensitive oscillator and a constant frequency one [16] .

The sensitivity of the whole system is about 2 ÷ 3
$\dfrac{gr/cm^2}{°C}$

5.3. Capillary tube sensor [17] [18]

It is mainly formed by a capillary tube where flows
the fluid at the temperature that has to be measured. Its operation is based on the fact that the value of the resistance varies
with the temperature.

If the resistance is laminar and neglecting the effects
of viscosity variation, one gets :

$$\frac{\Delta P_0}{Q} = \frac{T}{K} = R$$

where ΔP_0 is the pressure jump through the capillary tube,
Q the mass flow, T the fluid temperature, K a constant and
R the resistance.

Fig. 10 shows the scheme of a sensor of this type. The
capillary tube on the left side is at a temperature T , different
from T_0 which is the temperature of the right one.

The two capillary tubes are connected to the controls
of a proportional amplifier. As the resistance is different, the
flow in the capillary tubes is different and the amplifier gives
a pressure difference ΔP at its outputs.

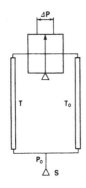

Fig. 10

Another way of utilizing a capillary tube sensor is to build a loop of resistances connected in the proper way. The resistances may be temperature sensitive or not. Temperature insensitive resistances are obtained with proper choice of material and design. It is then possible to have a loop in which only one resistance is temperature sensitive and to have an output formed by pressure differences which are proportional to the temperature [17].

5.4. Sensor with elastic moving member

It is possible to build temperature sensors which are based on the variation of size of solids and liquids.

Such a sensor which uses a fluidic element is shown in Fig. 11 [21].

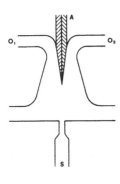

Fig. 11

It is a bimetallic strip sensor. The fluid with unknown temperature issuing from supply S strikes the splitter A , formed by a bimetallic strip, and is divided between the two outputs O_1 and O_2.

Variations of the flow temperature generate splitter deformations.

Owing to these deformations the flow distribution to the outputs changes, then a pressure difference which is proportional to the temperature follows.

Response time is about 150 m/sec. Sensivity is about $0,2 \dfrac{gr/cm^2}{°C}$.

Other elastic moving part sensors may be obtained with closing and opening nozzles [22] [23] .

6. Densitometer [1] [3]

Its operation principle is analogous to the one of the coflowing sensor. Its sizing is nevertheless different as it must be density sensitive and not velocity sensitive. The sketch of this sensor is shown in Fig. 12.

Fig. 12

The flow is issued from S into the ambient where there is the still fluid whose density has to be measured. The jet is then collected by receiving port 0 .

The jet entrains some ambient fluid and its slowing down depends on the density of the ambient fluid.

The collected pressure then decreases when density increases. This sensor may be used coupled with a reference unit, as seen for the coflowing sensor, such that only the density of the ambient fluid is different for the two units and the other parameters have always the same value.

REFERENCES

[1]   Tanney, J. W. : "Three fluidic sensors using un-
bounded turbulent jets"; 4th Cranfield Fluidics Conference, March 1970, British Hydromechanics Research Association, Cranfield,
England, paper R1;

[2]   Carbonaro, M. ; Colin, P. E. ; Olivari, D. : "The
deflection of a jet by a cross flowing stream
and its application to anemometry"; 4th Cranfield Fluidics Conference, March 1970, British Hydromechanics Research Association,
Cranfield, England, paper R2;

[3]   Briscoe, M. : "Some applications for fluidics in
oceanography"; 4th Cranfield Fluidics Conference, March 1970, British Hydromechanics
Research Association, Cranfield, England,
paper K2;

[4]   "Fluidics Components and Equipment 1968-69",
Pergamon Electronics Data Series, Pergamon
Press, 1968;

[5]   Foster, K. ; Cleife, P. J. : "The selection of a
fluidic transducer for sensing rotational speed
over a wide range"; 3rd Cranfield Fluidics
Conference, May 1968, British Hydromechanics Research Association, Cranfield, England,
paper E4;

[6]   Leathers, J. W. ; Davis, J. C. H. : " A new rotational speed sensor for fluidics"; 4th Cranfield Fluidics Conference, March 1970, British Hydromechanics Research Association,
Cranfield, England, paper T3;

[7]   Sarpkaya, T. : "A theorical and experimental in-
        vestigation of the vortex-sink angular rate
        sensor"; Fluid Amplification Symposium,
        October 1965, Harry Diamond Laboratories,
        Washington D. C. , vol. II;

[8]   Hellbaum, R. F. : "Flow studies in a vortex rate
        sensor"; Fluid Amplification Symposium,
        October 1965, Harry Diamond Laboratories,
        Washington D. C. , vol. II;

[9]   Kirshner, J. M. : "Fluid Amplifiers", McGraw Hill
        Book Co. , New York, 1966;

[10]   Sarpkaya, T. ; Goto, J. M. ; Kirshner, J. M. : "A
        theorical and experimental study of vortex
        rate gyro" from "Advances in Fluidics", The
        American Society of Mechanical Engineers,
        New York, 1967, pages 218-232;

[11]   Boydajieff, G. I.; Kasselman, J. T. ; Verge, K. Wm. :
        "Fluidics : a new tool for high temperature
        control systems", Fluidics Quarterly, Fluid
        Amplifier Associates, Ann Arbor, Michigan,
        vol. 1 n. 1, 1967;

[12]   Mayer, E. A. : "Other fluidic devices and basic
        circuits"; Fluidics Quarterly, Fluid Amplifi-
        er Associates, Ann Arbor, Michigan, vol. 1
        n. 2, January 1968;

[13]   Sarpkaya, T. : "The vortex valve and the angular
        rate sensor"; Fluidics Quarterly, Fluid Ampli-
        fier Associates, Ann Arbor, Michigan, vol. 1
        n. 3, April 1968;

[14]   Fish, V. R. ; Lance, G. M. : "An accelerometer for
        fluidic control systems"; The American Socie-
        ty of Mechanical Engineers", paper 67, WA/FE,
        29, 1967;

[15] Reeves, D.; Inglis, M.E.; Airey, L. : "The fluid oscillator as a temperature sensor", 1st Cranfield Fluidics Conference, September 1965, British Hydromechanics Research Association, Cranfield, England, paper D1;

[16] Otsap, B.A. : "Fluidics for ramjet control systems", Fluidics Quarterly, Fluid Amplifier Associates, Ann Arbor, Michigan, vol. 1 n. 4, July 1968;

[17] Walliser, G. : "Fluidic temperature sensor investigations for high gas temperatures" - "Fluidic Control Systems for Aerospace Propulsion", AGARDograph 135, September 1969;

[18] Johnson, E. G. : "Fluidic Gas Turbine Engine Controls" - Fluidic Control Systems for Aerospace Propulsion", AGARDograph 135, September 1969;

[19] Otsap, B. A. : "Fluidics in the control of advanced ramjet engines", "Fluidic Control Systems for Aerospace Propulsion", AGARDograph 135, September 1969;

[20] Kelley, L. R. : "A fluidic temperature control using frequency modulation and phase discrimination", Journal of Basic Engineering, June 1967;

[21] Torkelson, T. J.; Wilson, J. N.: "A fluid temperature sensor", I. F. A. C. Symposium on Fluidics", November 1968, Published by P. Peregrinus, Stevenage, England, paper C4;

[22] Novak, H. D. : "Application of a pure fluid amplifier in a simple fluidic cryostat", 3rd Cran-

field Fluidics Conference, May 1968, British Hydromechanics Research Association, Cran field, England, paper B 3;

[23]   Thivey, A. A. : "Les amplificateurs à fluide à déviation de jets", Sovcor Fluidique, Automatisme, Le Vésinet, France.

LIST OF FIGURES

Chapter 4

POWER ELEMENTS

1. Introduction

Power elements are not used to perform logic operations, but to control fluid currents having large flow and/or
high pressure.

In many cases they are the natural actuators of signals
coming from fluidic logic elements, and this is the case of momentum gain supersonic amplifiers.

In other cases they work as interface elements between
low pressure signals and mechanic pneumatic power members.
This is the case of pressure gain supersonic amplifiers.

In other cases their purpose is the control of large
fluid flows, gases or liquids. Elements of this kind are the diverter valve and the leg amplifier.

In any case the purpose of the power element is not to
perform a logic function but the control of a power fluid flow.
This control may simply be switching of the fluid flow into a
duct or into another one, or interruption of the passage (vortex
amplifier).

Several phenomena are exploited by power elements.
Separation and reattachment of a supersonic current from the
walls of a divergent channel are used in supersonic amplifiers.

Pressure recovery of a supersonic current which strikes a nozzle coaxial to emitter nozzle is used in the confined jet amplifier and in the vortex pressure amplifier. Several phenomena like momentum transfer, Coanda effect, contracted jet effect are used in the diverter valves and in semi-mechanical elements. The separation of a jet from a curved wall is used in leg amplifier. Finally, the vortex effect is used in vortex amplifiers.

2. Supersonic amplifiers

2.1. Operation principle and illustration of various kinds of supersonic amplifiers

These amplifiers are among the most important fluidic power elements. Their operation principle is the separation of a supersonic flow from the walls of a divergent channel and its reattachment [1] [2].

Supersonic amplifiers (see Fig. 1, 2, 5, 7) are provided with a convergent-divergent duct supplied by the power flow S. The hyperexpanded flow which flows in the divergent duct must overcome a positive pressure gradient (positive going on towards the outlet). If the geometry of the element and supply pressure have been chosen in the proper way, the flow separates from the wall forming some shock waves and reattaches on one side only, after forming a depression bubble.

This phenomenon is similar to wall effect ; the difference between the two is due to the fact that, for a given geometry, the bubble position depends on supply pressure. The rise of the bubble does not correspond, in fact, with the throat of the nozzle.

The consequence of this fact is that controls are placed rather downstream of the nozzle throat, in the wall of the divergent channel.

The control of a supersonic amplifier is similar to that one of wall effect elements and can be effected in two ways.

The first one is to cause an increase of bubble pressure by blowing from the control duct. Switching of the power jet is therefore due to a pressure control.

It is also possible to use the ambient pressure itself for controlling the element. In fact one can close one of the control ports leaving the other one open to atmosphere, then the jet switches to the side with closed control: the element works in closing control. The latter possibility is quite interesting for operation economy. In fact supersonic amplifiers work with high powers, and control flows, even though much smaller than power flows, may be quite considerable.

Of course, for closing control one needs valves or electrovalves with large enough passage areas, and this fact may limit the working frequency. In order to avoid that, one can use pressure control either with small valves or, better, with fluidic elements.

One should remember also the possibility of using, besides bistable and monostable types, amplifiers with flow evenly spreaded between the two outputs if controls are idle. This happens if the rate of flow can induce symmetrical separation from walls downstream of controls.

Supersonic amplifiers may be divided into two groups: momentum gain and pressure gain amplifiers.

In the first ones the high energy supersonic flow is switched to have a flow with a different momentum without re-transformation of kinetic energy into pressure. These elements are used in problems involving momentum propulsion.

The second ones are used for control of mechanical actuators and must have therefore a high pressure recovery .

2.2. Momentum gain supersonic amplifiers [3][4][5][6][7][8][9]

These amplifiers are often used in rocketry for thrust vector control. In this case they must work with gases which leave the combustion chamber.

Fluidic elements, which do not have moving members, are very suitable for control of these gases, which have a high temperature and contain abrasive particles.

In fact solid propellent gases contain solid particles of aluminium oxide, which generate troubles in each point of the wall they hit.

Supersonic elements may be digital (mainly bistable)
or proportional elements. But in proportional elements the
power jet hits the central splitter, which wears more quickly.

Therefore bistable elements are usually preferred.
Another cause of this preference is the higher gain of digital
amplifiers.

Gain of proportional amplifiers decreases when jet
deviation increases. Wall effect digital elements have an high
gain even when jet deviation is large.

Last but not least, digital amplifiers do not amplify
noise together with signal.

Supersonic amplifiers which work at 70 Kg/cm^2 pressure and 2900 ° C temperature have been built.

Operation time of such elements is limited by the
short life of the missile, and in many cases is 10 sec about.

Fig. 1 shows the sketch of a supersonic amplifier for
momentum control. No diffuser is used : the high speed flow is
directed to the outputs placed on the missile surface.

For these elements either bidimensional [5] or
tridimensional shapes have been used, the latter with circular
convergent-divergent nozzle and with a bidimensional body
placed in the surroundings of control ports and jet separation
[3] [5]. There are other cases which have circular
nozzle and circular output ducts [9] .

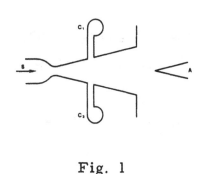

Fig. 1

In some particular cases axisym-
metrical geometries have been used
[8] . Materials used are usually
stainless alloys, while some parts
of the element that are less exposed
(like divergent walls downstream of
the controls) may be made with aluminium. These elements have been
used for reaction-jet control and for
rocket thrust vector control.

A typical application is missile attitude control. This
control must be made both on pitch and yaw planes. These
planes are perpendicular one another and their intersection is
the missile axis.

Let us assume that missile axis deviates, on one of
the two planes, from its correct attitude. To correct missile
attitude one should put a reaction-jet on the same plane and
giving a thrust opposite to missile deviation.

The reaction-jet intensity must be proportional to the
deviation. Therefore a proportional control is needed.

Control is obtained with digital elements by modulating
the time amplitude of each half wave [4] [6] [7] [8] .

If one puts two bistable amplifiers on pitch and yaw
planes respectively and makes them oscillate at a frequency
higher than the response frequency of the entire system (30 ÷

÷ 80 Hz), there is no effect if the time of any half-wave is constant.

An effect will be shown if the jet discharges for a longer time from an output than from the other one, at a constant oscillation frequency.

Fig. 2 shows the sketch of an asymmetric amplifier [4]

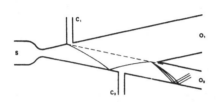

Fig. 2

The calibrating thrust is given by output O_2, activated by control C_1. When this thrust is unnecessary control C_2 activates output O_1, which gives a jet whose thrust helps the missile thrust.

The figure also shows the shock waves which are present in a bidimensional inviscid flow. The dashed line represents the free stream boundary.

For attitude control four of these elements are required, two on each plane.

Control may be obtained by closing control with electrovalves [3] or by small electrovalves which vary control flow [8] or by use of fluidic elements [6] [7].

In the latter case cylindrical vortex chambers are useful for impedance adjustment in connections between elements.

Control flows can be not larger than 10% of the power flow.

Fig. 3 shows an axisymmetric amplifier used for attitude control of a spin stabilized missile.

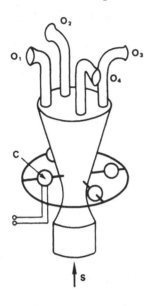

Fig. 3

It is an element with conical nozzle provided with four outputs $(0_1, 0_2, 0_3, 0_4)$ and four controls. The flow is deviated towards an output by a control jet activated by control electrovalves C. In this way one element has the same function like two or four elements seen before. Similar elements and techniques can be used also for missile roll control.

In Table I some characteristic data about supersonic amplifier operation are collected.

Fig. 4 (see next page) shows the sketch of a large fluidic valve [9]. The valve works by closing and opening controls (connecting them to atmosphere).

It is a kind of device intermediate between a supersonic amplifier and a diverter valve.

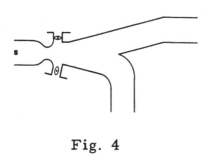

Fig. 4

Supply pressure is 3 atm only. The device can work in subsonic flow too. This valve, which has circular cross section ducts and nozzle, has been used for flow control in a VTOL aircraft.

2.3. Pressure gain supersonic amplifiers [10] [11] [12] [13]

These elements are useful to control pneumatic actuators. The signal utilized at the output is a pressure signal. For these elements an important problem is the sensitivity to output impedance, as they often work with blocked outputs.

Maximum recovered pressure is 25% ÷ 35% of supply pressure P_s .

Fig. 5 shows the shape of a supersonic bistable.

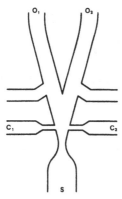

Fig. 5

The shape is similar to that one of a wall effect element with two vents and two outputs $0_1, 0_2$. The power nozzle, instead, has the characteristic convergent-divergent shape with control placed along the divergent walls.

Elements may be sensitive to output impedance or not. Not sensitive elements may become sensitive if the cross area of the control duct is too small.

Stability increases and sensitivity decreases if the control is made by blocking of a control duct.

Output external pressure affects the supersonic flow through the boundary layer. Then, more stable elements were studied with total discontinuity in the boundary layer, by means of a diffuser separated from the element [11].

The elements can be controlled also with a pressure signal.

Control characteristic has then the trend shown in Fig. 6.

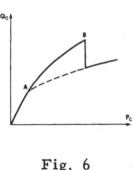

Fig. 6

Control pressure P_c and flow Q_c increase until the jet switches. B is the switching point if output are not blocked, A is the switching point when outputs are blocked [12]. Supply pressure p_s reaches $10 \div$ $\div 15$ kg/cm^2. Control flow needed for switching decreases when p_s increases.

Pressure gain may be high and in some cases, with blocked outputs, it may be even higher than 100.

Working frequencies reach 200 Hz.

Fig. 7 shows the sketch of a supersonic amplifier studied by NASA [13].

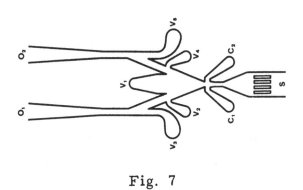

Fig. 7

Supply pressure is 4 kg/cm^2. Recovered pressure is very high and reaches 40% of p_s. The particular shape of output ducts and vents V_1, V_2, V_3, V_4, V_5 makes the element very stable and not sensitive to the output impedance.

Control pressure is 7% of p_s, and control flow is 6% of the power jet flow.

3. Confined-jet amplifier [14] [15]

This is a power element with very high pressure recovery. In some cases it can reach 97%. Even the working pressure can be very high, between 3 and 70 kg/cm^2.

This element behaves like a proportional one.

Element control is easily obtained by varying output flow from a proper chamber.

It can therefore be controlled by a vortex element as it will be seen later on.

Fig. 8 shows the sketch of the confined-jet amplifier. The element is formed by two cylindrical coaxial ducts placed inside an enclosure.

Supply jet S is collected by receiver 0. Output signal is picked up from 0 .

The element is controlled by adjusting the pressure in enclosure E. The vent H in the enclosure wall allows pressure adjustment either by varying the output flow, or by imposing a pressure by means of a pressure regulator.

The behaviour of the element depends on gap ℓ existing between ducts S and 0, and on diameter of ducts D .

Fig. 9 shows the trend of the curves which give the relationship between output pressure and enclosure pressure for different values of ratio ℓ/D. The plot shows output pressure P_0 (measured with blocked output) versus enclosure pressure P_E . Both pressure are normalized with supply pressure P_S .

Output and supply duct diameters are the same. One can see that the value of ℓ/D has a large influence on the trend of the curves.

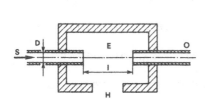

Fig. 8

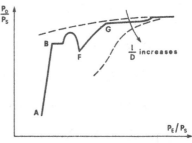

Fig. 9

If $\dfrac{\ell}{D} \simeq 2$ it is possible to distinguish three different modes of operation. In the first mode (A B) pressure gain

$G_P = \dfrac{\Delta P_0}{\Delta P_E}$ is the highest, about 7 and maximum recovered

pressure is 86%.

In the second mode (F G) pressure gain is 2 and recovered pressure reaches 96%.

In the third mode (beyond point G) recovered is even higher and linearity increases. Pressure gain, however, is very low, lower than 1.

Usually the behaviour of the first mode is the most favourable.

Output characteristics of the element are shown in Fig. 10. Supply pressure P_S is the same for all the curves.

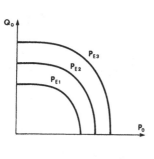

Fig. 10

Each curve has been traced for one value of P_E Increasing P_E output pressure and output flow increase. For a given working condition one can see that increasing output impedance, output flow Q_0 decreases and output pressure P_0 increases.

Fig. 11 shows a confined-jet amplifier controlled by a vortex amplifier. In this case the enclosure vent has been connected to the radial supply duct of the vortex.

Increasing control (C) pressure the intensity of the

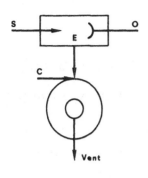

Fig. 11

vortex which is generated increases. Then output flow is reduced and P_E increases. If the vortex has the proper size it is possible to have a given P_E, independent of the flow which discharges from enclosure into vortex radial port, and corresponding to a given control pressure of the vortex.

Then, there is a direct relationship between vortex control pressure P_C and confined-jet amplifier P_E.

Another way to control the element with a vortex valve is shown in Fig. 12.

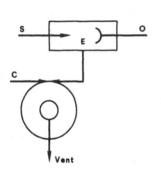

Fig. 12

The enclosure is now connected to a tangential vortex input whose cross section is quite larger than cross section of control C. The effect of a control flow increase is now a reduction of the vortex intensity. Then, increasing control flow the pressure loss through the vortex decreases. At the same time P_E decreases and the element does not show for a certain range of P_E values a significant reduction of control pressure.

In this range it is then possible to get very high pressure gains.

4. Vortex pressure amplifier [16]

This one is a proportional power element with high pressure recovery. It is quite fit for working with very high output impedances, there is with closed or quasi-closed output. These conditions are present, for instance, when the element controls mechanical actuators like pneumatic pistons.

In these cases the element has a high pressure gain.

The vortex pressure amplifier works with supply pressure from 1 to 50 kg/cm^2 or more.

The sketch of the element is shown in Fig. 13.

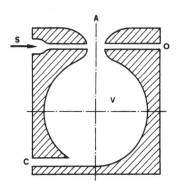

Fig. 13

Supply duct S and output duct O are coaxial and separated by a short gap. On a side of the ducts there is the vortex chamber V connected to external ambient A by means of a convergent-divergent duct. Control duct C inlet is placed tangent to the vortex chamber wall.

When control signal is absent the jet coming from S reaches the output duct with no pressure difference between vortex chamber and external ambient.

In such conditions the jet goes straight.

If output duct is blocked or quasi-blocked the fluid

flow is divided into three parts one of which discharges into
internal ambient, the other one goes into output O and the
third one into chamber V where it generates a vortex.

If one sends a control signal, then control flow contrasts the spinning of the vortex and the power jet is subject
to the pressure difference between vortex chamber and external ambient.

As the jet is deviated, then the power jet fraction
which enters into the chamber V decreases. In a short time a
new equilibrium condition is reached with deviated supply jet.

Output pressure decreases when power jet is deviated
and the jet deviation depends on control pressure.

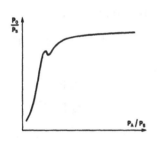

Fig. 14

Fig. 14 shows the trend of the output pressure P_0 versus ambient
pressure P_A of the element with
blocked output and absence of control pressure. Both pressures P_0
and P_A are normalized with supply pressure P_S .

Recovered pressure is rather high and is more than
80% of P_S when $P_A / P_S > 0,25$. When $\dfrac{P_A}{P_S}$ becomes lower
than 0.25 the recovered pressure decreases quickly.

It is convenient to work in the high recovery range of
the element.

This curve is similar to that of the confined-jet amplifier (see Fig. 9).

In both cases curves represent the recovered pressure
of a supersonic jet impinging a duct coaxial to the jet axis, with
blocked output.

Ambient pressure and enclosure pressure are, in both
cases, pressure at the nozzle exit.

However, control and working range of the two elements are quite different.

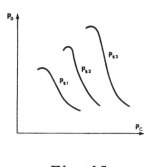

Fig. 15

Fig. 15 shows the trend of output pressure P_0 versus control pressure P_C.
The parameter of the curves is supply
pressure P_S. All the curves are traced
for the same constant value of P_A / P_S.
The same curves give the pressure
gain of the element $G_P = \dfrac{\Delta P_0}{\Delta P_C}$.

The gain is negative and varies between
6 and 10.

Flow gain is much lower; therefore, the use of vortex
pressure amplifier as flow amplifier is not recommended.

The sensitivity of the element to different output impedances is shown in Fig. 16 where output pressure and flow
(P_0 , Q_0) versus output impedance (z) are plotted.

Of course, when impedance increases, pressure increases and flow decreases.

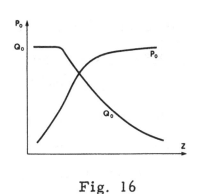

Fig. 16

The dynamic behaviour of the element is approximately the same of a linear system of the first order. Time constant of a vortex element with vortex-chamber diameter of 25 mm is 2 m/sec about. Frequency response tests showed that at a frequency of 100 Hz the decrease of signal amplitude is 4 db and phase shift is $60°$ about.

The element does not show hysteresis.

Repeatability is good; about 1%.

5. Diverter valves

Diverter valves are useful to distribute the flow among different channels.

They can work either with gaseous substances or with liquids.

Diverter valves have usually a big size, as they have to work with rather high flows. The flow is subsonic.

These devices are useful in chemical industry where they work with corrosive fluids, or in nuclear facilities where the presence of man has to be avoided as much as possible. As they have no moving part they do not show a quick wear and valve life depends on the wall resistance to corrosion.

A kind of diverter valve is shown in Fig. 17 [17].

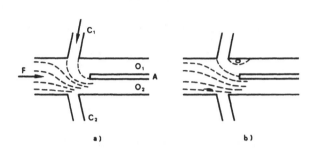

It is a bidimensional duct divided in two parts by splitter A The flow F coming from the left side flows symmetrically into the two ducts when control signals

Fig. 17

are absent.

The flow can be deviated into either output by means of the jet issuing from one of the control ports C_1 and C_2.

The deviation can be obtained in different manners. In Fig. 17 two extreme situations are shown.

In Fig. 17a the jet coming from control C_1 switches the flow F to output O_2 and there is no flow in the passive output O_1.

In Fig. 17b the main flow F only is switched to output O_2, while control flow flows through O_1.

In this case the energy required for control flow is lower.

Then, intermediate situations are possible between the two extreme ones.

Practically if, as usual, the flow is turbulent, there

is a mixing of control and power flow.

As a consequence one part of power flow will flow through the passive leg. In the case shown in Fig. 17a this phenomenon will be less important than in the case of Fig. 17b.

The element operation is based upon momentum principle.

The limits are due to the fact that the power required by the control jet is larger than flow F power.

Fig. 18 shows a diverter valve with "vena contracta" which can work with gases, liquids and with two different fluids, i. e. liquid for the power jet and air in the external ambient [18].

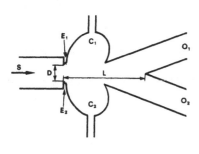

Fig. 18

The element is characterized by two edges E_1 and E_2, which are perpendicular to the direction of flow and are placed at the outlet of the power duct. Downstream of the edges there are two vortex chambers connected to control ducts.

The control is obtained by means of a pressure difference in control chambers C_1 and C_2. The jet is deviated towards the chamber where the pressure is lower and in this chamber a depression bubble is generated (Fig. 19b).

The phenomenon is similar to the one shown by wall effect elements; in the present case, however, it is easier to deviate the jet.

The main difference is the different direction of the jet at the outlet of the power jet duct.

In usual wall effect elements the power jet has the direction of the nozzle axis and the deviation begins downstream of the nozzle outlet.

In these elements the jet has a certain amount of deviation at the exit of the power nozzle.

The different behaviour is due to the blades E .

The phenomenon develops like follows : The blades produce a contraction of the jet, generating therefore inward velocity components perpendicular to the flow direction.

Static pressure at the center of the jet is then larger and the velocity smaller than the ones at the boundaries of the jet itself (Fig. 19a).

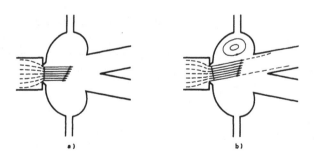

Fig. 19

Now a pressure difference downstream of the controls increases the curvature on the low pressure side of the "vena contracta" and reduces it on the high pressure side.

The trend of the velocity is shown in Fig. 19b, with the higher value on the low pressure side.

The increase of speed gives here a further decrease of the static pressure, while the opposite thing happens on the other side.

Then, the jet has a further deviation.

Summarizing, as a consequence of the jet contraction, a pressure difference downstream of the blades influences the pressure at the nozzle exit and therefore the jet has a certain amount of deviation already when issuing from the nozzle.

This element can be controlled by closing control or with control pressure.

The pressure recovery may be very high and depends on the ratio $\frac{L}{D}$ (Fig. 20) where L is the splitter distance from blades and D the nozzle width (Fig. 18).

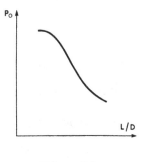

Fig. 20

In water the recovery reaches 95% when $\frac{L}{D}$ = 3. If the splitter is too near the blades the fluid flows from passive duct too. The element may work with high flows. With a 40 x 40 mm cross section nozzle the element works with flows from 200 to 2500 1/min of water. If vents are

downstream of the vortex chambers and if the element geometry
is proper, the behaviour of the element can be tristable, bistable or proportional.

6. Double leg elbow amplifier [19]

This one is a proportional amplifier which gives
large flow and power amplifications.

It works by exploiting both the separation of a flow
from a curved wall and the jet interaction with momentum transfer.

Fig. 21 shows the sketch of the amplifier.

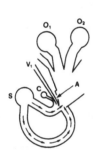

The flow issuing from supply S flows
in the two legs of the element (flow direction is indicated by the arrows). The
upper leg is called "active leg" because
the control works in this leg and the
lower leg is called "passive leg". The
control works as a disturbance of the
boundary layer of the current which
flows in the active leg.

Fig. 21

In this way, the separation point of the flow from the inner duct wall moves upstream and the velocity distribution
changes. This is the basic phenomenon for the element operation.

Fig. 22 shows an enlargement of the active leg with
velocity profiles in two different situations.

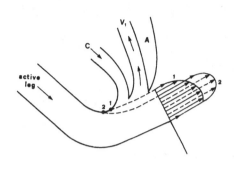

Fig. 22

The curvature of the duct is rather high. When control signal is absent, the current separates from inner wall at a point 1 which depends on duct configuration and on boundary layer conditions. The fluid is submitted to actions of inertia which force it against the outer wall.

The velocity profile at the duct exit is distorted and the zone of higher velocity is near the outer wall. The control signal modifies the boundary layer conditions and cause the flow separation in a point 2, upstream of point 1. The fluid is then forced with more effectiveness towards the outer wall and, as the flow is about the same while the cross section decreases, there is an increase of velocity with a variation of the momentum of the output current. The flow trap A eliminates the slowest part of the flow by deviating it into the vent V_1.

At the active leg output the fluid meets the passive leg current. A momentum transfer between the two currents follows. The final current goes to outputs O_1 and O_2 depending on the momentum of the active leg current.

Moreover, the flow - before reaching the receiving port - follows the curved wall of flow trap A for a certain length depending on the direction of the resulting jet that is on the control conditions. This element can give flow gains (up to 300) owing to boundary layer control and the other artifices. Pressure gains are rather low (up to 3).

The recovered pressure can reach 75% of supply pressure, which is very low ($7 \div 10$ cm $_{H_2O}$).

Power gains are very high (up to 500) as flow gains are high.

Noise is low as fluid velocities are low.

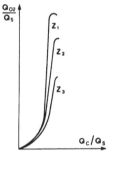

Fig. 23

Fig. 23 shows the trend of output Q_{O_2} versus control flow Q_C. The parameter Z is the output impedance. Higher flows are obtained with lower impedances. Output and control flow are normalized with supply flow. The curves show a good linearity.

The double leg elbow amplifier has been built as a small logic element and as a large power element (0.90 x 1.20 meters about) with power up to 50 CV.

Applications have been made to flow control in turbo-generator units.

7. Semimechanical fluidic elements

A particular group of power elements is the one composed by fluidic elements with a moving part.

In this way some of the main advantages of fluidic elements, like quick and sharp switching and a short response
time are preserved. At the same time some performances of
the element are improved, like the high energy loss and the
sensitivity to output impedance.

These problems are quite important for supersonic
elements.

From the point of view of reliability the semimechanical elements have very few moving members (usually one or
two) which only make small angular displacements.
The element shown in Fig. 24 uses a moving flapper F [20].

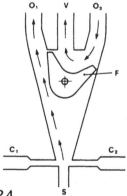

Fig. 24

The introduction of the flapper
makes the output O_1 independent
of O_2 and viceversa. Therefore,
the element works well even with
large pressure differences between active output and passive
output, while supersonic amplifiers can not work in such con-
ditions. The figure shows the flapper F which connects supply
S to output O_1, while output O_2 is connected to the central
output V .

This output is quite useful when the element works with oil in oleodynamic systems, where the utilized fluid must be carried into the reservoir.

The switching of the element is obtained by means of controls C_1 and C_2.

With this element it is possible to have a better efficiency with respect to elements without flapper because all the power flow goes to the active output. Pressure losses in the element and control pressure are lower.

The switching time is also lower.

These elements, even if designed for supersonic flows, can work with subsonic flows too (i. e. with lower pressure ratios).

Another kind semimechanical fluidic valve is shown in Fig. 25. It is a large element for large flows [9].

Fig. 25

Its power nozzle and ducts have a circular cross section. The element has no control duct and is provided with two butterfly valves placed in the output ducts. One of these valves is open and the other one is closed. The supply flow goes into one of the outputs or the other one depending on the valve position.

The operation of the element is based on the attachment

of a supersonic or subsonic flow to a wall. In Fig. 25 the flow is attached to the left wall; the left output is active. Valve A is open and valve B is closed.

The switching is obtained by exploiting the sensitivity to output impedance that is quite large with supersonic flows.

It should be pointed out that the element has no vents.

By closing the butterfly valve, the active duct pressure increases. When this pressure is high enough the jet becomes unstable and switches.

One gets the following advantages with respect to similar mechanical valves :

1) switching is instantaneous;

2) the force on each butterfly valve is lower. This is due to the fact that in mechanical valves the butterflies must hold all the thrust of the fluid, while in semimechanical fluidic elements the butterflies only have to make instable the jet attached to the wall. Therefore the force on each butterfly is about one half of the value they must hold in fully mechanical valves.

8. Vortex amplifiers [21] [22] [23]

These elements allow the regulation of a fluid flow within a large field of values.

They work then as variable resistance valves even if they have no moving part.

For this fact they are also called "vortex valves". They can work either with gas or liquids and can work well even if the fluid is not well filtered.

The principle of operation is that of the vortex. There are two kinds of elements, with vents and without vents.

Fig. 26 shows the sketch of a vortex amplifier without vents.

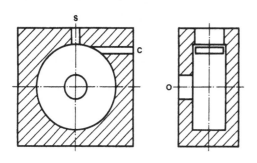

Fig. 26

The element is formed by a cylindrical chamber with a radial input S, a tangential control C and an axial output O. The supply flow is radial. More sophisticated models can have more than one supply port, or a uniform radial flow by supplying along the whole circumference of the chamber. With this purpose one could use also a porous ring.

When control is absent the flow goes directly to the output. The pressure in the chamber is constant and the flow is determined by the resistance due to the outlet.

The presence of a control signal generates a vortex. Inside the vortex neglecting viscosity effects, the tangential velocity increases.

The theory says that going towards the center of the

vortex, the increase of velocity is inversely proportional to the radius, while pressure decreases.

The more the tangential velocity at the outer edge of the vortex is high (i. e. the more control signal is strong), the more the effect is strong.

The consequence of radial pressure gradient in the vortex is that of a resistance increase and a flow reduction.

The trend of output flow Q_0, normalized with maximum output flow $Q_{0\,MAX}$, is shown in Fig. 28 versus $\dfrac{P_c - P_s}{P_s}$, where P_c is control pressure and P_s is supply pressure.

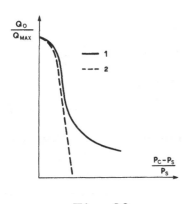

Fig. 28

The curve 1 refers to vortex element without vents. The flow Q_0 decreases quickly. The curves slightly different when P_s varies and in certain fields of operation they can be assumed independent of P_s. The output flow cannot be zero because the control flow is always present. Anyway, for a proper value of control pressure the supply flow can be zero, while for higher values of control pressure the supply flow changes its sign. That is a flow leaves the element through the supply port.

When supply flow is zero, control pressure is about

30 − 70% higher than supply pressure.

The output flow, which also includes control flow, can be reduced to 10 − 20% of its maximum value.

The vortex element is an element with negative gain which can give very high (until 100) flow gains $\left(\dfrac{\Delta Q_0}{\Delta Q_c}\right)$ [23].

It is possible to use many controls, summing or subtracting different signals.

In order to get a positive gain one should have a constant active control which generates a vortex opposite to the one generated by input signal.

The latter then reduces the vorticity and increases the output flow.

In order to nullify the output flow (see curve 2; Fig. 28), it is possible to use a vented amplifier.

In this device the output duct is placed at a certain distance from the exhaust orifice of the element, leaving therefore a vent **V** to atmosphere (Fig. 27).

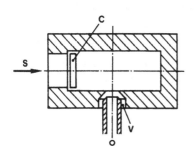

Fig. 27

When the vortex is absent, the fraction of flow that is collected by output duct depends on output impedance. If the vortex is present, the flow which issues from the vortex chamber has a conical shape. The cone angle increases with vortex intensity.

        A fraction of the flow is then dispersed into the extern-
al ambient.

        If the vortex has sufficient intensity, all the flow is
dispersed.

        The pressure recovered by a vented element depends
on many parameters, such as the area ratio between vent and
output duct cross sections and the shape of the exhaust orifice.

        The smaller is the summoned ratio, the higher is the
recovered pressure.

        This pressure can be very high in some cases and can
be higher than 90% of P_s, with blocked output.

        Supply pressure can be very high. With oil it is pos-
sible to get over 150 kg/cm^2.

        Fig. 29 shows the schema of two vortex valves connec-
ted to get a four way servovalve.

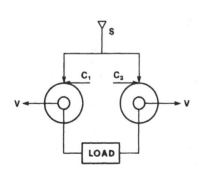

Fig. 29

The two elements are of the vented
type. Acting one of the controls, the
flow which crosses the corresponding element drops; therefore, a pressure difference on the load rises.
The vent allows the flow coming
from the load to exhaust into the
atmosphere.

Tab. I

Reference	Supply pressure P_s Kg/cm²	Temperature °C	Output thrust Kg	Operating time sec	Kind of element	Kind of control
[3]	63	Max. 1100	8	10	switch	control opening, with electromechanical flapper
[6] [7]	14-56	ambient	32 ($P_S = 50$ Kg/cm²)	–	digital amplifier; 2 outputs	fluidic
[8]	42-63	Max. 1400	20	10	digital amplifier; 4 outputs	control pressure, with selenoid valves

REFERENCES

[1]   Thompson, R. V. : "The switching of supersonic gas jets by atmospheric venting"; 2nd Cranfield Fluidics Conference, January 1967, British Hydromechanics Research Association, Cranfield, England, paper A5;

[2]   Shih, C. C. : "Flow characteristics in a supersonic fluid amplifier" from : "Advances in Fluidics"; The American Society of Mechanical Engineers, New Yor, 1967, pg. 129 and foll. ;

[3]   "Experimental Design of a Fluid Controlled Hot Gas Valve"; U. S. Army Missile Command, Redstone Arsenal, Alabama, Report N. RE-TR-62-9;

[4]   Harvey, D. W. ; Mc Rae, R. P. : "Steady Flow in a pure fluid valve TVC System"; Fluid Amplification Symposium, October 1965, Harry Diamond Laboratories, Washington D. C. , vol. IV;

[5]   Harvey, D. W. ; Mc Rae, R. P. : "Experimental Study of Fluid Controlled Valves"; Fluid Amplification Symposium, October 1965, Harry Diamond Laboratories, Washington D. C. , vol. IV;

[6]   Campagnuolo, C. J. ; Foxwell, J. E. ; Holmes, A.B. ; Sieracki, L. M. : "Application of Fluerics to Missile Attitude Control"; Fluid Amplification Symposium, October 1965, Harry Diamond Laboratories, Washington D. C. , vol. III;

[7]   Campagnuolo, C. J.; Holmes, A. B. : "Experi-
        mental analysis of digital flueric amplifiers
        for proportional thrust control"; 2nd Cran-
        field Fluidics Conference, January 1967,
        B. H. R. A., Cranfield, England, paper K3;

[8]   Holmes, A. B.; Foxwell, J. E. : "A development
        report on a fluid amplifier attitude control
        valve system"; Fluid Amplification Sympo-
        sium, October 1965, Harry Diamond Lab.,
        Washington D. C., vol. III;

[9]   Campagnuolo, C. J.; Holmes, A. B. : "A study
        of two experimental fluidic gas diverter
        valves"; 3rd Cranfield Fluidics Conference,
        May 1968, B. H. R. A., Cranfield, England,
        paper C1;

[10]   Bavagnoli, F. G. : "Experimental study on super-
        sonic fluid amplifiers"; 3rd Cranfield Fluid-
        ics Conference, May 1968, B. H. R. A., Cran-
        field, England, paper F8;

[11]   Thompson, R. V. : "Experiments relating to pres-
        sure recovery in supersonic bistable switch-
        es"; 3rd Cranfield Fluidics Conference, May
        1968, B. H. R. A., Cranfield, England, paper
        F10;

[12]   Bavagnoli, F. G. : "Performances of a Fluidic
        Power Switch"; X International Conference
        of Automation and Instrumentation, Milano
        1968, F. A. S. T. (Federazione Associazioni
        Scientifiche e Tecniche);

[13]   Griffin, W. S.; Cooley, W. C. : "Development of
        high speed Fluidic Logic Circuity for a novel
        pneumatic stepping motor" from "Advances
        in Fluidics"; The American Society of Me-
        chanical Engineers, N. Y. 1967, pag. 408;

[14]   Mayer, E.A. : "Confined jet amplifier"; Journal
         of Basic Engineering, A.S.M.E., March
         1968;

[15]   Mayer, E.A. : "Other fluidic devices and basic
         circuits"; Fluidics Quarterly vol. 1 n. 2,
         January 1968, Fluid Amplifier Associates,
         Ann Arbor, Michigan;

[16]   Otsap, B.A. : "The vortex Pressure Amplifier
         characteristics"; I.F.A.C., Symposium on
         Fluidics, November 1968, published by
         P. Peregrinus Ltd., Stevenage, England,
         paper B3;

[17]   Doctors, L.J. : "Fluid Jet Diverter Valve"; Me-
         chanical and Chemical Engineering Transac-
         tions, The Institution of Engineers, Australia,
         May 1967;

[18]   Barhton, S. : "A new type of fluidic diverting valve";
         4th Cranfield Fluidics Conference, May 1970,
         British Hydromechanics Research Association,
         Cranfield, England, paper A4;

[19]   "Fluidics"; Fluid Amplifier Associates, Ann Arbor,
         Michigan, 1965, pag. 31-32 and 201-204;

[20]   "Annual Report 1969"; Report FK 70503, Swedish
         Committee for Fluidics, Sweden;

[21]   "Fluidics Systems Design Guide"; Fluidonics Divi-
         sion of Imperial Eastman Corporation, Chicago,
         Illinois;

[22]   Taplin, L.B. : "Phenomenology of vortex flow and
         its application to signal amplification"; Fluid-
         ics Quarterly vol. 1 n. 2, January 1968, Fluid
         Amplifier Associates, Ann Arbor, Michigan;

[23]   Rivard, J. G. ; Walberer, J. C. : "A Fluid State
        Vortex Hydraulic Servovalve"; 21st Nation-
        al Conference on Fluid Power, Chicago,
        Illinois, October 1965.

LIST OF FIGURES

Chapter 5

INTERFACE ELEMENTS

1. Introduction

Interface elements are used as input and output elements in many fluidic systems. Generally speaking, input interface elements are necessary when input pulses are not fluidic. In this case we have non-fluidic sensors or we must transform another logic (say, electric, electronic, ...) into a fluidic logic.

Output interface elements are often used to convey control signals to power elements such as pneumatic or oleodynamic cylinders. In this case the presence of interface elements is mainly required by the different level of energy of the signals : low level for fluidic elements, high level for power elements. Interface elements are necessary for connection to oleodynamic systems, as different fluids are used at different energy levels. In many applications interface elements are preferred to fluidic power elements because the power consumption of the latter is very high.

In other cases, output interface elements are used to translate a pneumatic-fluidic signal into an electric signal.

The interface element will then be used either if it has to operate directly an electric power element, such as a motor, or if it has to provide the input of a low-power electrical logic system.

Interface elements are therefore important either for connection to power organs or for translating signals to a different kind of logic.

The latter case has a particular interest in order to have electrical-fluidic systems which may exploit the advantages of each technique.

Input interface elements are electric-to-fluidic transducers. Many kinds of fluidic sensors can be considered as mechanical-fluidic interface elements (proximity sensors, interruptible jet, ...).

In some particular cases high pressure-low pressure input interface elements may be required.

Output interface elements are fluidic-to-electric elements both of high power and low power type.

Other output interface elements are fluidic-to-pneumatic and fluidic-to-oleodynamic elements.

2. Input interface elements : electrofluidic interface

2.1. Micro-electric valves

A very simple way to get a fluidic signal from an electric one is to use a micro-electric valve.

This method allows a direct control on the fluid by

means of an electrically moved mechanical member. This member simply opens or closes some ducts and does not give amplification to the signal.

On the contrary, other types of interface elements
("flapper-nozzle" systems) amplificate the signal. These elements will be examined later on.

Among the various kinds of electrovalves it is better
to use the ones which do not use spools, which need lubricated
air. In fact, fluidic elements work with dry air.

The sketch of a normally open three-way electrovalves
is shown in Fig. 1.

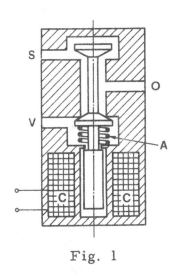

The central body of the valve is
made of ferromagnetic material.
When the coil C is excited, the
central body is attracted downward
therefore connecting output O to
vent V. The spring A contrasts the
solenoid action and pushes the valve
upward. Therefore, in absence of
an electrical signal, the output is
connected to supply S .

Fig. 1

Usually valves of this kind work with pressures up to
$7 \div 10 \text{ kg/cm}^2$ and their size may be very small.

Fig. 2 (see next page) shows another kind of three-way
valve whose supply may be connected to the two outputs $(O_1 \& O_2)$

one at a time (°).   Moving member is now simply formed by

a metallic disc D which can have

an axial displacement in a cylin-

drical chamber. Disc displace-

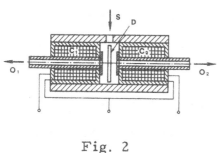

ment may be to left side if coil

C₁ is excited and to right side if

coil C₂ is excited. In any case, a

coil excitation leads to close the

Fig. 2

output duct placed on that side of

the element and to open the oppo-

site output.

In order to get a good sealing of the element, rubber

gaskets may be used between disc and chamber walls.

With such electrovalves rather high frequencies were

reached until 200 cps [1].

Another interesting kind of solenoid valve with a very

small moving mass is shown in Fig. 3 [2] and it is also used

for proportional operation.

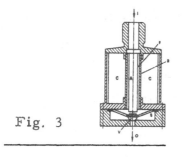

Solenoid core A is placed in the

center and is surrounded by the

ducts for fluid F, which flows

from input I to output O. A cylinder

Fig. 3

B, bounds the core and is the ex-

(°) Fig. 1 valve also may realize the same function if O be-
    comes supply and S and V the outputs.

ternal boundary of the flowing duct of the fluid.

A spring S forces the valve ball V to close the output
0.

The spring is provided with some vents in order to let
the fluid flow.

When solenoid C is excited, its core attracts the valve
ball. A passage for the fluid is therefore open. Passage area
depends on the equilibrium between spring and solenoid forces.

By an appropriate control of electric current it is possible to have a proportional adjustment.

2.2. Wall heating element [3]

It is a fully no moving part interface element. The
principle of operation is the control of the separation point of
a jet from a curved wall by varying wall temperature.

The sketch of this element is shown in Fig. 4.

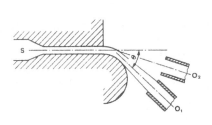

Fig. 4

From supply S through a
duct with rectangular cross
section the jet discharges
to atmosphere. At the outlet section one of the duct
boundaries is interrupted
and the other one continues
with a curved wall. The
jet is attached to the wall

and separates from it at a point downstream, then flowing with an angle Θ by respect to the duct axis.

The separation angle depends on geometry, Reynolds number and wall temperature. For a given geometry, when Reynolds number **Re** increases the angle Θ increases gradually and reaches a discontinuity point where it shows a sudden increase (Fig. 5).

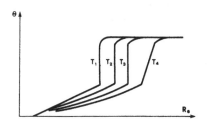

Fig. 5

At this point there is a laminar separation of the jet, then a reattachment followed by a turbulent separation. Beyond this point the separation angle Θ is practically independent of Reynolds number.

By decreasing **Re** the element may show an hysteresis depending on the geometry of the element itself.

By increasing the wall temperature the curve moves to right side and, at a constant Reynolds number, Θ decreases (Fig. 5). Angles and temperatures are directly proportional quantities in a certain range of Reynolds number and temperature values.

It is therefore possible to operate either in a proportional way, working in the summoned range, or in a digital way. In the latter case one must take care that the temperature increase (at a constant Reynolds number) causes the decreasing

of angle Θ from asymptotic value to a very small one. A receiv-

ing duct O_2 is placed to receive the jet when it is in the second

configuration.

In order to generate the temperature rise one may,

for instance, put on the curved wall a metallic ribbon crossed

by electric current. The response time of the element depends

mainly on the time necessary to ribbon heating.

2.3. Electric discharge element [4] , [5] , [6]

This electrofluidic element is a no moving part ele-

ment. The element has two couples of electrodes which can give

an electric discharge. Then, two phenomena are possible :

generation of shock waves and generation of expansion waves

due to the rise of temperature depending on the heating induced

by the discharge.

If these disturbances are sent to the power jet of a

fluidic amplifier it is possible to control the element.

Many electrode arrangements are possible. In the bi-

stable of Fig. 6 electrodes are placed in two control chambers

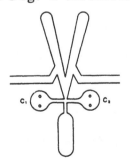

Fig. 6

C_1 and C_2 [4] [6] . Discharge ef-

fects work then like pressure con-

trol in a common fluidic bistable.

In the bistable shown in Fig. 7

electrodes were placed downstream

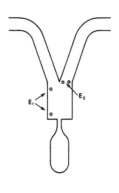

Fig. 7

of the power nozzle $[5]$. On the left side of the element, electrodes E_1 are placed in the point of jet attachment, on the right side of the element electrodes E_2 are placed at the mouthpiece of the output duct. For operation of electrodes E_2, a smaller amount of energy is required than for electrodes E_1.

In both cases, however, the discharge on a side switches the jet to the other side. Switching is obtained by adjustment of discharge time and intensity. With very short discharge times ($< 5\,\mu\,sec$) shock wave seems to be the more important phenomenon, while with longer times the thermic expansion is more important.

With an electrode arrangement like the one shown in Fig. 6 it is possible to control a proportional amplifier by adjusting intensity and frequency of electric pulses $[4]$.

With elements of the kind shown in Fig. 6, both ar - rangements with two electrodes and with three electrodes have been used. The third electrode is then used to prime discharge between the two main electrodes.

The possibility of obtaining a different behavior of the same element has been observed.

When the discharge strikes in a control chamber it is possible to get the switching of the jet to the opposite side and a further discharge between the same electrodes has no effect. The former is the normal operation with deviation control.

It may happen that the second discharge switches the jet to the initial position, the third discharge may switch again the jet, and so on. The element then behaves like a binary counter. That is, it switches each time a discharge takes place, and switching is independent of the couple of electrodes which is activated and of the jet position.

Another thing may happen: the jet may switch towards the side of the element where the discharge has taken place. The element behaves like a bistable with closing control.

The operation fields in which the element has different behaviours depend on supply pressure, on output impedances and on discharge modality [6] .

Response time, with a power nozzle 1 mm wide, is about 1 m sec. This kind of interface element has several troubles.

First of all it is necessary to work with relatively high voltages (200 ÷ 400 V). Secondly, positioning and adjustment of the electrodes must be very accurate. Last but not least, the problem due to electrode wear, even if they are made with very resisting materials like tungsten, which consequently gives metalization of control chambers.

2.4. Electrofluidic element controlled by vibrations and pressure pulses

Another way of having interface elements is to control fluidic elements by means of vibrations. A proper source of vibrations is able to generate in the surrounding air a lot of disturbances, pressure waves, rarefaction waves, and acoustic pulses. If these disturbances act in a correct manner on the jet of a fluidic element, the fluidic element itself can be controlled [7]. Some possible solutions of such a control applied to monostable and bistable elements will be examined.

A part of the problem is to have the proper vibration generator. With this purpose one could use electromagnetic generators, but a good solution is offered by the use of piezoelectric materials. The resulting elements are quasi-static elements (there is only an elastic vibrating wall).

Other advantages and drawbacks of piezoelectric elements by respect to electromagnetic ones, will be examined in the part reserved to electrofluidic elements with closing control. Now let us examine some fundamental properties of piezoelectric materials [8] [9].

Piezoelectricity can be defined through its effects : it is an electric polarization produced by mechanical stresses in a certain class of crystals. Polarization is proportional to the stress intensity and changes its sign together with the stress. On the contrary, an electric polarization produces a mechanical stress and the correspondent deformation.

Well known piezoelectric materials are quartz crystals.

tals.

Some ceramic materials also have very strong piezoelectric properties. Their behavior is however different and, in order to behave like single piezoelectric crystals, they must have a permanent polarization.

Among these substances there are the barium titanate $(BaTiO_3)$ and the lead zirconate titanate.

The former are the piezoelectric materials more fit for electrofluidic applications.

The most interesting piezoelectric effect is the production of a deformation due to the application of an electric field. This effect is well shown by a thin plate of piezoelectric material.

Let us consider a plate whose corners e, L, d, are respectively parallel to the axes of reference x (electric axis), y (mechanical axis), z (optical axis) (Fig. 8).

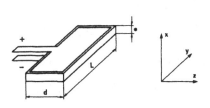

Fig. 8

Now let us apply an electric field having the same direction of axis x generated by a potential difference between upper and lower face of the plate. The latter will then show a dilatation (or a contraction) in the direction of axis x, and a contrac-

tion (or dilatation, depending on polarization sign) in the direction of axis y . There is no effect in direction of axis z . These
axes cannot be chosen at random. For instance, if the material
is quartz, axis z is the axis of maximum symmetry of the crystal, and x and y depend on the lying of crystal faces.

There are also other different modes of the plate deformation depending on the material. With quartz, however, the
described modes are the only possible and are the ones of great-
est sensibility with barium titanate and lead zirconate. If one
applies an alternate voltage to the faces of the plate, contrac-
tions and dilatations follow. Amplitude is usually very small and
higher values are obtained when excitation frequency coincides
with mechanical resonance frequency of the plate.

Therefore this device can be a good high frequency
vibration generator. In the element of Fig. 9 [10] , one of
these piezoelectric vibrators P, controlled by electric control
E , is applied to the wall of a proper fluidic element.

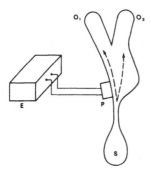

If the vibrator is not excited the
laminar supply jet S is directed
to output O_2 . The vibrations induce the transition of the jet to
turbulence. Then, a larger air
flow is entrained and the jet is
attracted toward the left wall
(Coanda effect). Active output is

Fig. 9

O_1 . If vibrations stop, the initial jet configuration takes place.

Fig. 10 shows the scheme of a bistable controlled with vibrations.

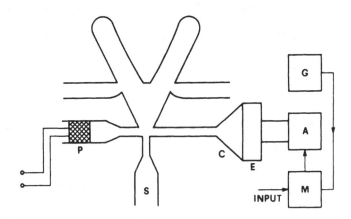

Fig. 10

The same control system can be applied to monostable elements. Two different solutions are shown. On the left side there is a control system with piezoelectric element [9] while on the right side a control with an electromagnetic element is shown.

The latter is formed by an earpiece E with a convergent cone C to collect and convey the sound.

The earpiece is controlled by a high frequency (5-8 kHz) wave modulation obtained from a power amplifier A, a modulator M and a frequency generator G [11] [12]. Control pulses (INPUT) are applied to the modulator M.

It is also possible to control these elements with single pulses [12] [13] .

The purpose of the generator is, in this case, to generate a pressure signal in a small volume.

This purpose is reached if the diaphragm is the wall
of a chamber such that when it moves the volume decreases and
pressure increases.

The electric control system is now very simplified.
Control voltages are about 10 V or less and therefore quite
consistent with electronic systems.

If the earpiece is connected to a bistable control and
earpiece input is a square wave, the bistable element switches
when the electric signal appears and switches again when the
signal becomes zero.

If earpiece input is formed by slow decay pulses, like
the ones due to a condenser discharge, the element shows mem-
ory. In order to increase the sensitivity of the system without
building special elements one can utilize the method shown in
Fig. 11.

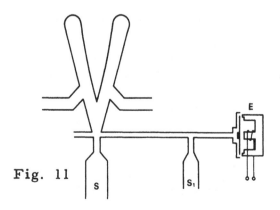

Fig. 11

On the line which connects
the earpiece E to the fluidic element an additional
supply has been placed
(S_1). Then, a D. C. signal is always present and
the amplitude of the re-

quired control signal is reduced.

A bistable element with such a control system reached a switching time of 1, 2 m sec. and maximum frequency of
750 Hz. Minimum control power is about 10 mW.

Last but not least, there is the possibility of controlling directly with a vibration generator a turbulence amplifier,
as a laminar jet is very sensitive to sound [14] [15].

2.5. Variable geometry element

This element is formed by a fluidic element which has
a moving elastic part whose deformation switches the power
jet. This kind of element is represented by the piezoelectric
nozzle element [16].

A bistable nozzle is built with thin plates of a piezoelectric material (B_1 and B_2 , Fig. 12).

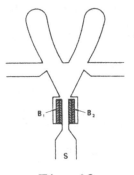

When the plates are excited, their deformation is large enough to permit the
jet deviation to either of the outputs.
However, the required deformations are
usually larger than the simple contraction or dilatation of a single plate. To increase the wall displacement are used

Fig. 12

bimorph or multimorph plates (two or more plates joined together). In these multimorph plates any single plate is pre-
coated with a thin metal layer, usually silver. Then, the

plates are joined together.

Electric connexions may be in series or in parallel (Fig. 13). The latter need half the voltage and twice the charges with respect to the former.

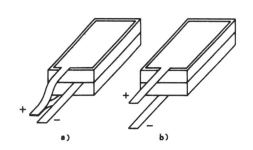

Fig. 13

Depending on the displacement of any single plate, when excited it is possible to have flexural or torsional deformations of the multimorph plate [9]. In the case of the interface element with piezoelectric nozzle a bimorph plate with flexural deformation is needed.

Another kind of variable geometry interface is the oscillating jet element [17].

This device is often used for electric-to-oleodynamic conversion but it is fit for fluidic systems working with air too.

The sketch is shown in Fig. 14. Nozzle N is revolving around pin A and is supplied with fluid from supply S, by means of an elastic tube. The jet emitted from the nozzle is collected by two receiving ports placed in front of the nozzle itself. Distance between nozzle and collectors is at least two times the nozzle diameter.

The connection of the nozzle to the rod B, made with ferromagnetic material and carrying two coils, is rigid.

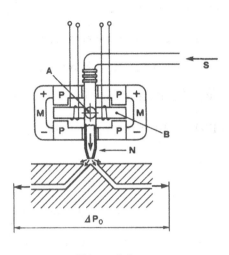

Fig. 14

The rod is immersed in a magnetic field generated by one or two permanent magnets M and by pole pieces P made with ferromagnetic material. The whole is the electromagnetic control system of the nozzle. The coil excitation in fact produces the rod magnetization.

Then, attraction and repulsion forces rise (depending on the sign) with polar pieces P. Rod and nozzle rotation follow.

The use of two coils allows nozzle displacement towards right or left side.

If the jet is symmetrical with respect to the receiving ports when coils are unexcited, the pressure recovered by each port is the same. If, on the contrary, the jet is deviated, the recovered pressure will be different.

This fact is analogous to the behavior of a proportional amplifier with jet deviation. Now, however, the deviation is due to an electric signal.

The pressure difference ΔP_0 obtained in this way can be used to control further proportional or digital elements. If one wants to use the oscillating jet element with proportional systems, feedbacks of different kinds (mechanical, electrical, etc.) can be used to increase the field of linearity.

For oleodynamic systems this element is sometimes preferred to the flapper-nozzle one, which will be shown later on, because internal ducts do not have throttled sections. This is important in order to avoid bad operation due to fouling.

2.6. Closing and opening control element

Another kind of interface element works by closing and opening the control ports of a fluidic element. This technique can be applied both to monostable and bistable elements, if these elements are such that it is possible to deviate the power jet by closing and opening control ports.

Therefore, it is necessary to put some mechanical members near the outlet of control ducts and move or warp these members by means of an electric control.

Of course, these mechanical members can be very small and light, as their purpose is only to permit or not the communication of a control port with the external ambient. The electromechanical control can be either of the electromagnetic type [13] [18] or of the piezoelectric type [19] [20].

The distance between the outlet of control ducts and the mechanical member which causes the switching of the element, can always be very little. In some commercial fluidic elements, which may be controlled with closing control (Aviation Electric), this distance must be less than 0. 04 mm to have an effect on switching.

Fig. 15 shows the scheme of a bistable controlled by two piezoelectric plates which close the control ports.

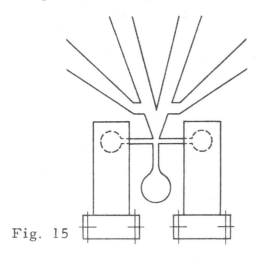

The two control ducts communicate with the external ambient through two circular ports on the upper part of the element. Bimorph plates are placed on the ports. These plates (Fig. 16) are restrained at an end and they normally close the control port. When

Fig. 15

the plates are deflected by means of electrical excitation, the

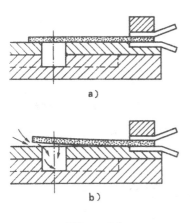

a)

b)

Fig. 16

control ports are open and external air can be entrained from atmosphere into the element. This air entrained into the separation bubble destroys the bubble itself and causes the switching of the jet. This kind of control is consistent with electronic systems and can be incorporated in integrated circuits.

Minimum control voltages are 45 V about with bimorh in parallel. The frequency can reach 450 Hz about. This frequency is influenced by the eigenfrequency of the plates and by the time necessary to reach the flow required for control.

Fig. 17 shows the scheme of a back pressure element controlled with control port closing by a microrelay.

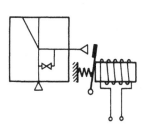

Fig. 17

This forms a very simple and economic system. Maximum frequencies are 30 Hz about [13] . The choice of an electromagnetic or piezoelectric system can also be made with the help of the following considerations :

1) piezolectric systems have moving elastic parts only while electromagnetic ones use moving parts;

2) piezoelectric actuators are insensitive to possible disturbances due to external magnetic fields caused by iron masses placed in the neighbourhood of the element;

3) piezoelectric materials are rather brittle;

4) piezoelectric materials can work with low temperatures only. Beyond a certain temperature piezoelectric properties disappear.

This temperature is 100° C for barium titanate and 550°C for quartz.

2.7. Electric control on a flapper-nozzle system

A convenient way to get an electrofluidic interface is

to act on a proximity sensor. As a sensor a double nozzle system can be used (flapper-nozzle system). It receives and am-
plifies the signal and gives output signals high enough to control either static fluidic elements or membrane elements or
spool-valves.

Fig. 18 shows a double nozzle system controlled by a
relay which moves a flapper.

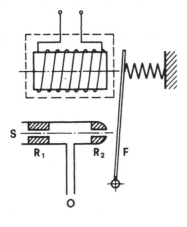

Fig. 18

When the exhaust duct through a
resistance R_2 is free the output
(0) pressure is minimum. When
flapper F approaches, it closes
the exhaust to atmosphere of the
air coming from supply S . Then,
output pressure increases.   The
system is basically a proportional one, but it can be used as a
digital one.

The output signal can control directly a fluidic element. With a correct shaping of supply resistance R_1 and exhaust resistance R_2, it is possible to have (when exhaust impedance is zero) an output signal negative, zero or positive, depending on what one needs.

By utilizing an earpiece as closing element, response
times of about 2m sec were obtained by controlling a fluidic
element [13]. As a closing element one can use of course a

piezoelectric system too [9].

In Fig. 19 the system is applied to the control of a double membrane element.

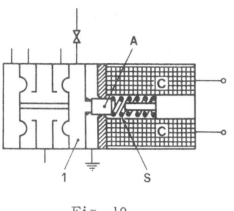

When coil C is excited, the member A is attracted to the right side against the spring action and chamber 1 discharges into atmosphere. The large pressure decrease in chamber 1 acts as a control signal on the double membrane element. The output signal which is equal to the supply pressure is about 1.4

Fig. 19

kg/cm^2.

The last system, shown in Fig. 20, is the scheme of a double flapper-nozzle system used to move a spool valve [21].

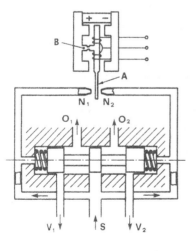

Fig. 20

The flapper control is electric. The system is quite smitable when the working fluid is oil. This is a well known electro-to-oleodynamic element and it may be a useful interface in oil-working fluidic systems. As it is shown in Fig. 20, the rod A is placed between the nozzles N_1 and N_2. The rod is movable because it is connected to the frame through a flexure bearing B .

The control is obtained with an electromagnetic system with a magnet M and pole pieces, as shown in the moving
nozzle element.

The system can work either as a proportional or as a
digital one.

Closing of one nozzle causes a pressure increase in
the control chamber of the spool-valve.

The spool-valve shown in the figure is the three-posi-
tion type. In the central position, corresponding to the central
position of the rod, outputs O_1 and O_2 of the spool-valve are
disconnected both from supply and from exhaust.

3. Input interface elements : high pressure - low pressure interface

3.1. Introduction

Pneumatic-to-fluidic interface does not require,
generally speaking, any particular element because the signal is transmitted with the same fluid. It is only necessary
to reduce the energy level, for instance by means of a resistance connected in series to the fluidic element, or by
means of a resistance connected like the former and another
one connected in parallel and which discharges to the external
ambient (ground).

Of course the cleanliness of the air used, with possible presence of oil traces, has to be proper for the fluidic elements. Otherwise it is necessary to have proper elements like in the case that high pressure fluid is oil.

3.2. Disc interface and bellows interface

A high pressure-low pressure moving disc interface is shown in Fig. 21 [22].

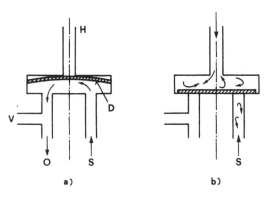

Fig. 21

A moving disc D is contained in a cylindrical chamber. At an end of the chamber there is the high pressure input H. At the other end there is the low pressure supply S and low pressure output 0. In absence of the high pressure, the signal passes from S to 0 (Fig. 21a). This passage is closed when the pressure in H is higher than pressure S. The valve has the vent V connected to the output.

In this valve, after the switching there is no passage of fluid from H to 0 or S. The danger of soiling of the fluidic circuit air is therefore very small.

In the case that high pressure fluid is oil, and low pressure fluid is air, it is possible to use mechanical-pneumatic systems. One can use for instance a flapper-nozzle system, supplying the nozzle with air and taking the low pressure signal from the nozzle. To close the nozzle it will be used a moving system with a bellows, or a small piston, or a membrane moved by the high pressure oil. In such a way, it is possible to get a complete separation between the two fluids.

4. Output interface elements : fluidic-to-electric interface

4.1. Introduction

These interface elements can be used either to control power members, like electric motors, or to translate fluidic logic into electronic logic.

Generally speaking, in the first case the problem of response time does not exist and the solution can be very simple making use of microswitch systems or magnetic circuit systems. In this case there are moving parts.

In the second case bigger problems arise, mainly concerning the response time.

4.2. Microswith elements [23] [24]

The scheme of an interface of this kind is shown in Fig.22 (see next page). The fluidic pressure signal coming from input I acts the membrane M against the action applied by the spring S. The piston P moves together with the membrane and acts the microswitch W opening or closing an

electric contact. It is also possible to use slide valve systems
or rolling membrane systems [24].

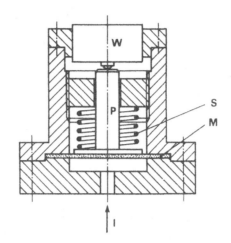

Fig. 22

4.3. Magnetic circuit element

Also opening and closing of a magnetic circuit - obtained with a pneumatic-mechanic system - can be used to have an
electric output signal [25].

The scheme of such an interface is shown in Fig. 23.

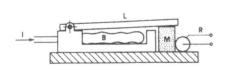

Fig. 23

A reed switch R , with normally open contact, is kept closed
by magnet M . The lever L is
made of magnetic material.
The operation of the device
is as follows. When in the
bladder B there is no pressure
the magnet is shunted by the lever and the reed switch is open.

A fluidic signal coming from input I inflates the bladder. Therefore, the lever rotates and the magnetic shunt is removed. The reed switch then closes the contact.

4.4. Interface for electronic circuits

There are many types of fluidic-to-electric interface fit for the scope. Let us mention all kinds of transducers (piezoelectric, inductive, capacitive, potentiometric ones, etc.) used in current instrumentation. These transducers are both of analog and digital type.

As fluidic-to-electronic interface we can remember, in particular, two types of interface : the one with piezoelectric plate and the pressure sensitive transistor. The latter seem to have good possibilities of development and application. Fig. 24 shows the sketch of a piezoelectric interface [12].

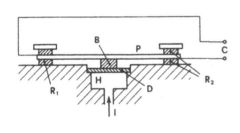

Fig. 24

A plate P of piezoelectric material is supported at its ends by small rubber blocks R_1 and R_2. The center of the plate is connected to the diaphragm D through the rubber block B. This diaphragm is one of the walls of a chamber H to which the pressure is sent.

The diaphragm itself can be stuck to its support by epoxy casting.

The deformation of the plate, due to the pressure in the chamber H, is detectable at the contacts C .

5. Output interface elements : fluidic-to-pneumatic and fluidic-to-hydraulic interfaces

5.1. Introduction

These elements allow the conversion from low pressures of fluidic systems (from few tens of cm H_2O to few tenths of atmosphere) to the pressures used in pneumatic systems ($5 \div 10$ kg/cm^2) and to the high pressures of oleodynamic systems. These elements are very often used in industrial applications.

There are different kinds of interface, also because the air used in the high pressure circuit may contain oil traces or not.

The first case is the most common one and it always occurs when in the power circuit there are pneumatic pistons or slide valves which need a continuous lubrification for a corrects operation.

All output interface elements fluidic-to-pneumatic and fluidic-to-hydraulic contain mechanical moving or elastic members.

This happens because with this device one wants to stop the fluid flow without consuming air, while with fluidic power elements this does not happen.

Another difference is that the maximum pressure recovery in the power circuit can reach 100%.

These elements, in fact, do not exploit - for their
operation - dynamic phenomena which cause energy losses.

5.2. Interface elements with large control area

These ones are elements in which the small available
pressure acts on a rather large surface, generating therefore
a control force which is high enough [23] .

This force is utilized to move mechanical members
which, closing and opening some ports, switch in the correct
way the air of the power circuit. These moving members can
be spool-valves, or spear-valves, or other moving members.

In the first case the control force must overcome
mainly the friction due to the gaskets which contrast the spool
displacement. In fact the power circuit air does not have an action if the spools are balanced.

In order to get the balancing of the spool, the thrust
given by the high pressure air must be the same in the two
directions of displacement axis. This happens if in any one
of the high pressure chambers of the slide valve, the areas of
the surfaces lapped by air on the sides are equal. The area is
measured on the section perpendicular to the axis.

In the second case the control force win the thrust
that high pressure air applies to the spear or other moving

members. Then, to have the control, the power fluid must act
on a surface smaller than the one on which low pressure air
acts.

Spool-valve elements are usually able to work with
lubricated air on the power circuit.

In fact sealings are usually made with lubricated
0-Rings. Fig. 25 shows a three-way balanced spool-valve with
spring reversal.

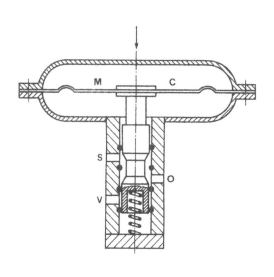

Fig. 25

Output 0 is alternatively
connected to supply S or
to exhaust V . Control is
obtained by sending air
into the control chamber
C , whose lower wall is
formed by the membrane
M . A rod connects M to
the spool. Instead of the
membrane one could use
a rigid disc which moves
axially together with the
spool. Then the control chamber seal is sliding and can work
without lubrication if a proper choice of the gasket material
has been done.

Fig. 26 shows the sketch of a four-way spool-valve,
controlled by two rolling membranes M [24]. One gets a

good seal and small forces resisting to displacement in control

chambers. Of course, there is friction in the internal spool

seals.

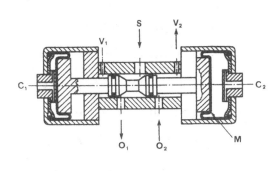

Fig. 26

In the configuration shown in Fig. 26 output O_1 is connected to supply S and output O_2 to exhaust V_2. By increasing pressure at control C_1 the connections are : S to O_2 and O_1 to V_1.

Minimum control pressures are $0.2 \div 0.3$ kg/cm^2. Power air pressures go until 10 kg/cm^2 and more, depending on the slide valve. Response times are quite longer than for other types of interface elements and depend very much on filling time of control chambers.

In spite of that it is possible to reach response times of the order ot $20 \div 30$ m sec. [26] .

The same scheme can be used also for oleodynamic slide valves [24] .

These interfaces need power air with oil traces. Oil is not necessary if the moving member is a spear or a pin. This one can also be of the type shown with the electrovalve of Fig. 1.

Fig. 27 (see next page) shows the scheme of a low

pressure-high pressure element with a pin : it is the "Pressure-Area Amplifier" [23]. This one is a two-way valve. High pressure supply S can be connected to output O or not.

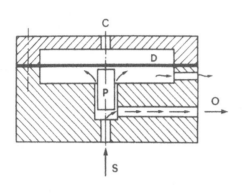

The connection is provided by pin P which can close more or less the output port. The pin is pushed up by the high pressure air. The upper base of the pin touches a diaphragm D strained by the control signal C

Fig. 27

The pin equilibrium depends on the areas of the diaphragm and of the pin itself and on control and supply pressures.

It is possible to have a proportional adjustment as the output flow can vary from a maximum to zero.

The element can also work with dry air as the pin is air-lubricated.

This fact causes of course a small air loss through the pin itself. Another loss, very small indeed, is possible from supply to output when the element is closed.

Working frequencies of this element are rather high, higher than 50 Hz. Control pressure varies between 0 and 0,7 kg/cm^2 with a power pressure of 7 kg/cm^2.

5.3. Flapper-nozzle elements [13][23] [27] [28] [29] [30]

This method gives the best results. The system is the one seen for electro-fluidic flapper-nozzle interface elements. Now, however, closing and opening of the nozzle is obtained with a membrane strained by control pressure.

Like in already mentioned valves with large control area, also in this case there are mechanical moving or elastic parts, which open and close passage channels of high pressure air.

But now mechanical members are not directly moved by low pressure air of the control but by high pressure air of the double-nozzle system. Therefore large membranes and large control chambers are no longer necessary.

In other words, flapper-nozzle system works as an amplifying system, intermediate between low pressure of the control and moving members. Sensitivities are then larger and working frequencies higher.

Fig. 28 (see next page) shows the sketch of such an element.

Supply air from S is discharged into the external ambient through the double nozzle system (N_1 and N_2) and its outlet is on the lower side of control diaphragm.

Application of a signal C causes N_2 closing. The body of the valve is then pushed down and output 0 is connected to supply (Fig. 28b).

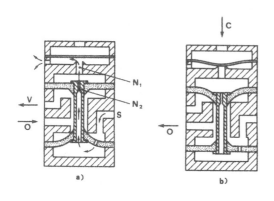

Fig. 28

These elements work well either when the power air is lubricated either when it is dry. A low pressure-high pressure interface proper for control of a double membrane element can be built in the same way as the valve shown in Fig. 19.

Now, however, the nozzle is open and closed by a membrane strained by the control [29].

Minimum control pressures are $10 \div 20$ cm$_{H_2O}$ about. Output pressures reach 10 kg/cm^2. Working frequencies up to 50 Hz.

Fluidic-to-oleodynamic interfaces have also been built following the same principle [30].

It is also possible to build systems where the flapper-nozzle acts on a high pressure spool-valve by means of a disc valve [22].

5.4. 0-Ring amplifier [31]

This is a particular low pressure-high pressure valve. It is formed by two parts : a three way valve controlled with

the spindle A (see right side of Fig. 29) and a body containing
the two membranes which control the spindle. Its operation is
like follows.

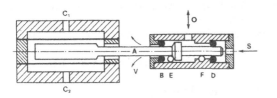

Fig. 29

If no control is applied to
the controls C_1 and C_2,
the spindle is in a central
position. 0-Rings B and
D give a good seal and
air does not pass through
them. When control C_2 is
activated the spindle is
pushed down and rotates around pin E. The radial displacement
on OR B is small and sealing is still good. Displacement on
OR D is large and air passes through the OR itself going from
supply S to output 0.

If control C_1 is activated the spindle moves upward by
rotation around the pin F. Now air passes through OR B. Then
output is connected to vent V.

The spindle collar limits the axial displacement of
the spindle, otherwise it could be pushed away by the pressure
existing inside the valve.

Power-circuit pressure from 5 to 10 kg/cm^2.

Minimum control pressure 0.03 kg/cm^2.

Comparatively high working frequencies.

5.5. Frictionless spool-valve and rotating distributor valve

Other kinds of low pressure-high pressure interfaces are based on the use of frictionless spool valves [20] [32].

In these elements spool and sleeve are made with thermally treated stainless steel and lapped with very high precision. There is no sliding gaskets and seal is given by the narrow passage between spool and sleeve.

The air itself forms a very thin film around the spool and lubricates the surfaces.

Then there is practically no friction and the forces needed for displacement are very small.

Interface spool-valves without friction work like usual spool-valves used in pneumatic and oleodynamic circuits.

The difference is that frictionless valves have no gaskets and are air lubricated. This fact permits to work well even with small control pressures without making use of large surfaces.

These interfaces work with dry air in the power circuit.

Control pressures are about 0.07 kg/cm^2, power circuit pressures are 3.5 kg/cm^2 about.

The maximum working frequencies that have been reached are 10 Hz [32].

Fig. 30 shows the scheme of the last (for this treatise) kind of interface : the rotating distributor valve [23].

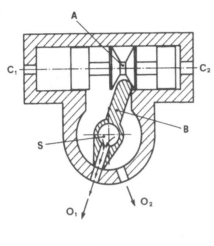

Fig. 30

The spool A is displaced by low pressure control air (C_1 and C_2 are control ports). The spool has, in its central part, a double collar which is engaged to a shaped arm of the rotating distributor B. The latter is so shaped that supply duct S is parallel to its axis of rotation (in Fig. 30 the axis is perpendicular to the plane of the sheet). Output duct is radial.

A rotation of distributor B corresponds to a translation of the spool, bringing the output duct of the distributor to cope with O_1 or with O_2.

It can therefore work as a diverter-valve or, changing the power connexions, as a three-way valve.

These interfaces work with lubricated power air and with pressure to 10 kg/cm^2.

Minimum control pressure is 0.14 Kg/cm^2.

Working frequencies are higher than 30 Hz.

REFERENCES

[1] Bell, R. W. : "An application of fluid logic to tele-
graphy as a research vehicle"; 1st Cranfield
Fluidics Conference, 1965, B. H. R. A., Cranfield, England, paper E1;

[2] Anon. : "Solenoid valve has one moving part";
Control Engineering, January 1968, pag. 39;

[3] McGlaughlin, D. W. and Taft, C. K. : "Fluidic elec-
trofluid converter"; Journal of Basic Engineering, June 1967, pag. 334;

[4] U. S. patent n. 3 263 695;

[5] Nyström, K. S. and Brodin, G. : "Control of a
boundary layer fluidistor by means of disrup-
tive discharge"; 1st Cranfield Fluidics Conference, 1965, B. H. R. A., Cranfield, England,
paper B4;

[6] Bistagnino, C. and Clauser, G. L. : "Studio speri-
mentale di un elemento elettrofluidico"; Atti
del X Convegno Internazionale Automazione e
Strumentazione, Milano 1968, F. A. S. T.;

[7] Gottron, R. N. : "Acoustic Control of Pneumatic
Digital Amplifiers"; Proc. of Fluid Amplification Symposium, May 1964, Harry Diamond
Laboratories, Washington D. C., Vol. 1; pag.
279;

[8] Cady, W. G. : "Piezoelectricity"; McGraw Hill
Book Co., New York 1946;

[9] Lee, S. Y. : "Piezoelectric actuators for fluid control applications"; Engineering Seminar, Penn-
sylvania State University, 5-16 July 1965, Engineering Proc. P45, pag. 100 and following;

[10]   U.S. patent n. 3 269 419;

[11]   Benson, C.A.R. and Hawgood, D. : "Dynamic
        acoustic switching of digital pneumatic am-
        plifiers"; 1st Cranfield Fluidics Conference,
        1965, B.H.R.A., Cranfield, England, Dis-
        cussion to paper B4;

[12]   Hawgood, D. : "Electrical transducers for fluid-
        ic systems in computer peripherals"; 2nd
        Cranfield Fluidics Conference, B.H.R.A.,
        Cranfield, England, paper F3;

[13]   Retallik, D.A. and Goedecke, H. : "Peripheral
        and interface devices"; 4th Cranfield Fluid-
        ics Conference, 1970, B.H.R.A., Cranfield,
        England, paper C4;

[14]   Auger, R.N. : "Turbulence Amplifier Design and
        Application"; Proc. Fluid Amplification Sym-
        posium, October 1962, Harry Diamond Lab.,
        Washington D.C., vol. 1, pag. 375;

[15]   Piantà, P.G.L. : "Effect on acoustic environ-
        ments on a fluidic turbulence amplifier";
        PR 68-227, Von Karman Inst., Rhôde St.
        Genèse, Belgium, June 1968;

[16]   U.S. patent n. 3 266 511;

[17]   Morse, A.C. : "Electrohydraulic servomechan-
        isms"; McGraw Hill Book Co., New York
        1963, pag. 37 and following;

[18]   Masuelli, G. : "Generatore di segnali penumatici
        a frequenza variabile"; Graduation thesis,
        Istituto di Meccanica Applicata, Politecnico
        di Torino, 1968;

[19] Render, A. B. : "An electrical control for fluid logic elements"; 2nd Cranfield Fluidics Conference, 1967, B. H. R. A. , Cranfield, England, paper F1;

[20] Render, A. B. : "Power outputs from fluidic devices"; 3rd Cranfield Fluidics Conference, 1968, B. H. R. A. , Cranfield, England, paper K9;

[21] Morse, A. C. : "Electrohydraulic servomechanisms"; McGraw Hill Book Co. , New York 1963, pag. 26 and following;

[22] Brewin, G. M. : "The disc or free foil valve as a practical logic element"; 3rd Cranfield Fluidics Conference, 1968, B. H. R. A. , Cranfield, England, paper E6;

[23] Dummer, G. W. A. and Mackenzie Robertson, J. : "Fluidic Components and Equipment 1968-69"; Pergamon Press 1969;

[24] Monge, M. : "Some industrial applications of the Moduflog systems"; 3rd Cranfield Fluidics Conference, 1967, B. H. R. A. , Cranfield, England, paper J4;

[25] Anon. : "Composite pressure sensors combine several technologies"; Product Engineering, October 1968, pag. 42;

[26] Belforte, G. : "Low frequency fluidic oscillators"; 3rd Cranfield Fluidics Conference, B. H. R. A. , Cranfield, England, paper T4;

[27] Hodge, J. and Hutchinson, J. G. : "Turbulence Amplifiers-Principles and Applications"; 1st Cranfield Fluidics Conference, 1965, B. H. R. A. , Cranfield, England, paper F2;

[28]   Wiesner, H. J. and Rüdle, M. A. : "Fluid devices
        for machine controls"; 1st Cranfield Fluidics
        Conference, 1965, B. H. R. A. , Cranfield,
        England, paper G2;

[29]   Köning, G. R. : "Design of inputs and outputs of
        digital pneumatic jet components for adapta-
        tion to the standardised pressure range";
        2nd Cranfield Fluidic Conference, 1967,
        B. H. R. A. , Cranfield, England, paper B5;

[30]   Ward, E. J. and Kemble, J. E. and Wheeler, K. J. :
        "A lathe control system incorporating fluid-
        ics"; 3rd Cranfield Fluidics Conference, 1968,
        B. H. R. A. , Cranfield, England, paper A4;

[31]   Lichtarowicz, A. : "0-Ring seal as a valve ele-
        ment"; 3rd Cranfield Fluidics Conference,
        1968, B. H. R. A. , Cranfield, England, paper
        E1.

[32]   Anon. "Numatrol Handbook NH-66"-Austin S.
        Beech and Co. Ltd., Leighton Buzzard,
        England.

LIST OF FIGURES

Fig. 1   - Three-way electrovalve.

Fig. 2   - Metal disc electrovalve.

Fig. 3   - Ball electrovalve.

Fig. 4   - Wall heating element.

Fig. 5   - Separation angle Θ versus Reynolds number for different wall temperature $T \cdot T_1 < T_2 < T_3 < T_4$.

Fig. 6   - Electric discharge element with control chambers.

Fig. 7   - Electric discharge element without control chambers.

Fig. 8   - Thin plate of piezoelectric material.

Fig. 9   - Monostable element with piezoelectric vibrator.

Fig.10   - Bistable controlled with vibrations.

Fig.11   - High sensitivity bistable-earpice element.

Fig.12   - Piezoelectric nozzle element.

Fig.13   - Electric connexions of a bimorph - a) in parallel b) in series.

Fig.14   - Oscillating jet element.

Fig.15   - Piezoelectric plates controlled element.

Fig.16   - Piezoelectric plates : a) control port closed - b) control port opened.

Fig.17   - Back pressure element controlled with a microrelay.

Fig.18   - Flapper-nozzle electrofluidic interface.

Fig.19   - Double membrane element controlled by a flapper-nozzle system.

GENERAL CONTENTS

Chapter 3. Metrologic applications

Printed in the United States
By Bookmasters